UNITED STATES COMBAT AIRCREW SURVIVAL EQUIPMENT

WORLD WAR II TO THE PRESENT
A REFERENCE GUIDE FOR COLLECTORS

MICHAEL S. BREUNINGER

Schiffer Military/Aviation History
Atglen, PA

Edited by Alan R. Wise.

All photos are by the author except as noted.

Cover photos are courtesy of David F. Brown.

Library of Congress Catalog Number: 93-90752

Printed in the United States of America.
ISBN: 0-88740-791-9

We are interested in hearing from authors with book ideas on related topics.

Published by Schiffer Publishing Ltd.
77 Lower Valley Road
Atglen, PA 19310
Please write for a free catalog.
This book may be purchased from the publisher.
Please include $2.95 postage.
Try your bookstore first.

DEDICATION

Kay - For her hard work and the long hours she spent helping me complete this book

- and especially her love and understanding

TABLE OF CONTENTS

	PAGE
PREFACE	7
ACKNOWLEDGEMENT	8
INTRODUCTION	9
DISCLAIMER	10
CHAPTER 1 SURVIVAL VESTS AND LIFE PRESERVERS	11
C-1 EMERGENCY SUSTENANCE VEST	12
E-1 RADIO CARRIER VEST	16
SL-1 SURVIVAL LEGGINGS	17
SV-1 SURVIVAL KIT VEST	18
SV-2 SURVIVAL VEST	20
CMU-24/P SURVIVAL VEST	26
SRU-19/P SURVIVAL CHAPS	27
SRU-20/P SURVIVAL VEST	30
SRU-21/P SURVIVAL VEST	32
SRU-32/P FRAGMENTATION SURVIVAL VEST	36
FIGHTER ATTACK VEST	37
ARMY LIGHTWEIGHT INDIVIDUAL SURVIVAL KIT	38
OV-1 AIRCRAFT SURVIVAL VEST	42
U.S. ARMY SARVIP VEST	45
WORLD WAR II LIFE VESTS	47
MK-2 LIFE VEST	50
MK-3C LP	53
LPA-1 AND LPU-23/P LP'S	54
MA-2, LPU-2/P, LPU-3/P, LPU-9/P, AND LPU-10/P LP'S	56
CHAPTER 2 SURVIVAL SEAT AND BACK PAD KITS	59
AN-R-2 RAFT KIT	60
C-2 RAFT KIT	62
C-2A RAFT KIT	64
PK-1 PARARAFT KIT	67
M-592 BACK PAD KIT	69
JUNGLE EMERGENCY PARACHUTE BACK PAD KIT	73
B-1 ALASKAN EMERGENCY PARACHUTE BACK PAD KIT	74
B-2 JUNGLE EMERGENCY BACK PAD KIT	75
B-4 EMERGENCY PARACHUTE KIT	78
PK-2 PARARAFT KIT	81
CNU-1/P SUSTENANCE KIT CONTAINER	83
ML-3 INDIVIDUAL LONG RANGE SURVIVAL KIT CONTAINER	85
ML-4 RAFT KIT	89
INDIVIDUAL OVERWATER, HOT AND COLD CLIMATE SURVIVAL KITS	91
EJECTION SEAT SURVIVAL KITS	97
MD-1 SEAT KIT	108
MB-2 SEAT KIT	110
U.S.A.F. SEAT KIT COMPONENTS	111
B-58 ESCAPE CAPSULE	114
OV-1 EJECTION SEAT KITS	118

TABLE OF CONTENTS

	PAGE
CHAPTER 3 PERSONAL SURVIVAL KITS AND FIRST AID KITS	**122**
F-1 EMERGENCY SUSTENANCE FLYERS CASE	123
E-3 AND E-3A 'PERSONAL AIDS' KITS	124
E-17 'PERSONAL AIDS' KIT	126
PSK-2 PERSONAL SURVIVAL KIT	128
1 AND 2- PART INDIVIDUAL SURVIVAL KITS	129
SEEK 1 KIT	133
SEEK 2, SRU-31/P, AIRMAN'S INDIVIDUAL SURVIVAL KITS	135
'TAC' KITS	138
HOT-WET ENVIRONMENT INDIVIDUAL SURVIVAL KIT	141
SRU-16/P MINIMUM SURVIVAL KIT	143
WORLD WAR II FIRST AID KITS, PERSONAL TYPE	145
POST WORLD WAR II FIRST AID KITS, PERSONAL TYPE	147
SNAKE BITE AND INSECT STING KITS	152
CHAPTER 4 SELECTED SURVIVAL COMPONENTS	**153**
SURVIVAL RADIOS:	
AN/URC-4	154
AN/URC-11	155
AN/URC-68 AND AN-PRC-17	156
AN/URC-64 AND ACR/RT-10	157
RADIO BEACONS, RT-20A AND AN/PRC-93	158
AN/PRC-63 AND RT-60	159
AN/PRC-90 AND AN/PRC-90-1	160
AN/PRC-112	161
STROBES AND SIGNAL LIGHTS	162
MK-1, MK-13, MK-124 SMOKE AND ILLUMINATION FLARES	163
FLARE LAUNCHERS	165
SIGNAL MIRRORS	168
RATIONS	170
MISCELLANEOUS FOOD PROCUREMENT ITEMS	173
DESALTING KITS	174
FIRE STARTERS AND MATCH CASES	175
SURVIVAL RIFLES	177
SURVIVAL TOOLS AND KNIVES	180
COMPASSES	183
SURVIVAL MANUALS	184
PERSONNEL LOWERING DEVICES (PLD'S)	185
ONE-MAN LIFE RAFTS	186
BLOOD CHITS, ESCAPE MAPS, PHRASE BOOKS	187
APPENDIX	190
REFERENCES	199

PREFACE

The topic of survival equipment involves a tremendous amount of information due to its various applications. Survival situations occur in peacetime and war, and in all climates and regions of the world. Many types of survival kits and accessories exist for the individuals, aircraft and ships that may need them.

Since before WWII, aircraft personnel have been issued or have had access to many types of survival equipment. This book is concerned with only that survival equipment actually worn or carried by an individual airman or crewmember, although many of the components are used in larger kits.

I have further chosen to concentrate on only the survival vests, seat and back pad kits, personal and first aid kits used since WWII. Parachutes, anti-exposure suits, flightsuits, flight helmets and escape maps, etc. are considered items of individual survival but require their own book because of the vast amount of information they can present.

The information and photographs in this book are all gained from items of personal collections, available military survival and equipment manuals, and conversations with persons who have used the equipment. Though I've attempted to be as thorough as possible, there are still items of which I am unaware of or do not have access.

I hope to provide as much information as possible to help current and future collectors enjoy, learn and hopefully exchange new information on the subject.

Acknowledgements

I wish to give special thanks to:

Scott Novzen for the illustrations on the chapter introduction pages, and who influenced my collecting of survival equipment.

Alan Wise for his editing skills and convincing me that writing this book was possible.

Gene Kaplan for his support and assistance.

I also wish to thank the following individuals and organizations whose contributions made this book possible:

Kym Aklan
Frank Davidson
Dave Davis
Robert Chad LeBeau
J.W. Peklo
John Scott
Kevin Smith
Andy Wear
Chris Woodul
Motorola Electronics
Natick Army Research Center

INTRODUCTION

Several guidelines are important in collecting survival equipment as well as in using this book.

While manuals may indicate specific components or arrangements of components in a vest or kit, this can sometimes be the exception rather than the rule.

The phases found in many manuals 'mission dictates' or 'at the local commanders discretion' can change what and how items are used. Also, many base, field and personal modifications exist that change the item. As components are improved or changed, they replace older items in the vests and kits.

Therefore, while the photos in this book depict specific components or arrangements, other variations can and will exist.

Component dates play an important role in survival equipment, not only for the year of manufacture and issue but as guidelines for determining expiration and inspection dates.

Early dated items can be found in later dated kits, i.e., a 1967 dated pocket knife in a 1980's survival vest. This is acceptable in certain components that do not have a limited shelf life or are prone to spoilage such as flares and medical supplies.

For example, a 1985 dated flare in a Vietnam era seat kit does not make a true Vietnam issue kit.

Each and every survival item has a useful life span until replaced by something better. Sometimes the most nebulous area is determining when a particular item came into service and when it was removed. Just when you think you've pinned down the specifics on an item, someone or some document will refute your previous findings.

Any dates indicated in this book are based on current information and may vary plus or minus until more concrete data is found.

A specific military branch is indicated as the user for many of the survival vests, kits, etc. in this book. Although not listed, other branches of the armed forces could use the same equipment. For example, the Marine Corp and Coast Guard would use many Navy items; and the Army and Air Force have similarity in their equipment.

Many tags and labels were faded or too illegible to photograph. To maintain consistency and legibility, all have been typeset and have been represented as close as possible to the original. Also, realize that although a specific manufacturer or label arrangement is shown, others can exist.

I welcome any opinions and comments, especially information that can clarify or add to the subject.

DISCLAIMER

All items depicted or described in this book are intended for collectors and display use only. Actual use of the items is not implied nor recommended.

CHAPTER 1

SURVIVAL VESTS AND LIFE PRESERVERS

The U.S. military use of survival vests began during World War II with the issuance of the C-1 Emergency Sustenance Vest. Various forms of life preservers (LP's) have been in use long before the survival vests and are used alone or with various life raft kits. Both survival vests and LP's continue to be major components of the U.S. military survival equipment inventory. Many types and variations have been developed by each service who have their own particular ideas of what is required to sustain their aircrews. While some items are common between the services, the majority of survival vests, contents, and LP's are unique to its specific branch of service.

The C-1 vest was employed during and after World War II, into Korea and even into Vietnam. This was due in major part to its availability and its usefulness. The many pockets could accommodate any number of items issued at the time, and few other vests were available in any quantities. There was a proliferation of survival vests types in the 1950's and 60's and other related garments being developed by the Air Force and Navy along with many individually designed and produced vests. However, it wasn't until the introduction of the Air Force SRU-21/P and the Navy's SV-2 series vests that the services became somewhat satisfied. Improvements continued and new ideas are still being tested today, as witnessed by the Army's SARVIP survival vest.

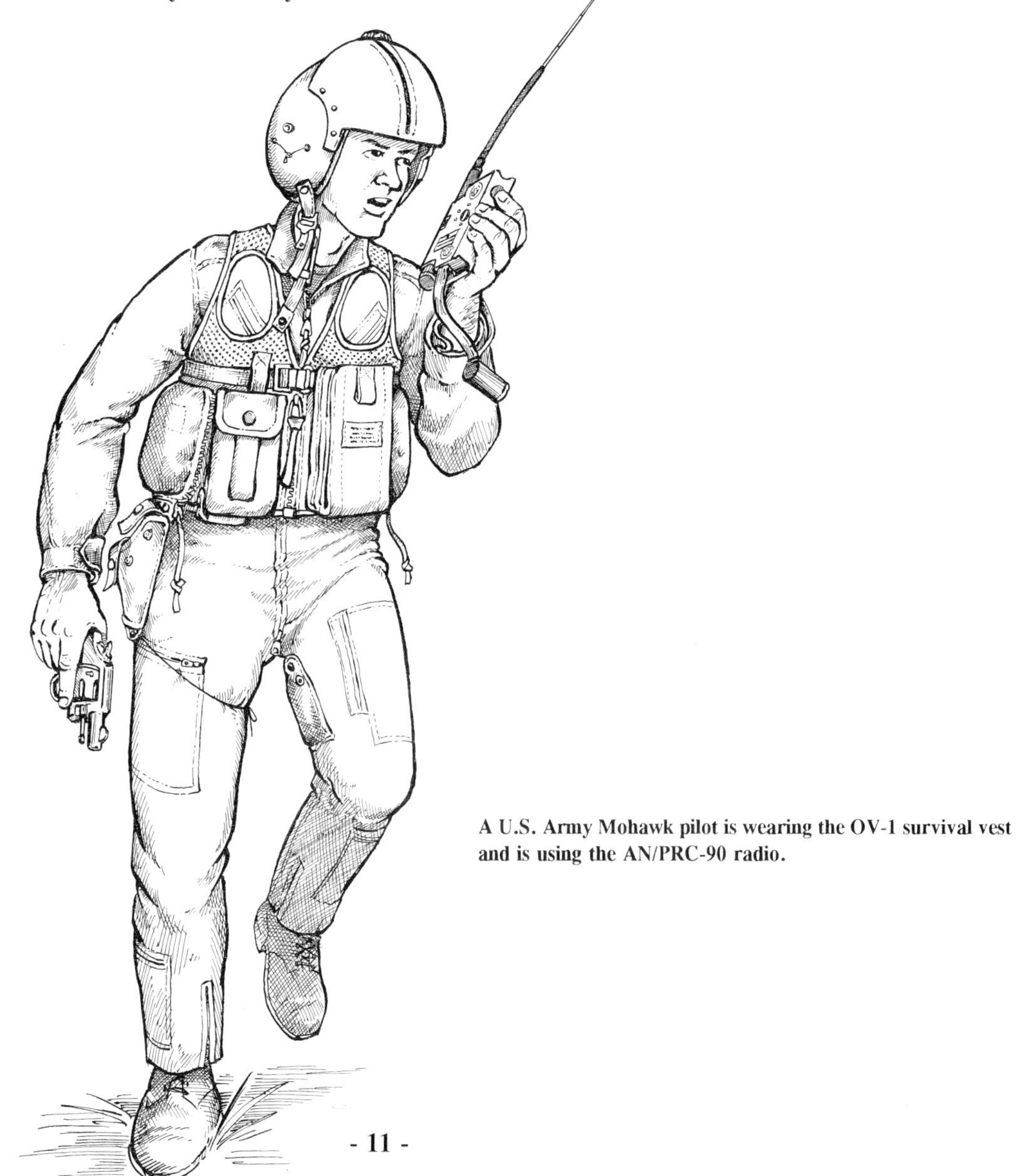

A U.S. Army Mohawk pilot is wearing the OV-1 survival vest and is using the AN/PRC-90 radio.

C-1 EMERGENCY SUSTENANCE VEST

C-1 vest (later pocket variation)

C-1 vest (back)

LABEL

VEST,EMERGENCY SUSTENANCE
TYPE C-1
SPEC. NO. 3206
ORDER NO. W33-038-AC6640
SEARS, ROEBUCK AND CO.
PHILADELPHIA,PA
MFD. BY RELIANCE MFG. CO. CHICAGO, ILL.
PROPERTY A.F. U.S.ARMY

Developed during World War II as the first survival vest for use by the Army Air Force, the C-1 Emergency Sustenance Vest has seen use by both the Air Force and Navy through Korea and even into Vietnam, although in small numbers. It was made of dark olive green tackle twill and weighed about 11 pounds when outfitted. There were three large buttons in front for closure. The vest was used alone or in conjunction with various seat and back pad survival kits which were also developed during World War II. While it was designed for use in all parts of the world, the C-1 proved better suited for the tropics than the artic regions.

The C-1 was issued one size but was adjustable for fit through three ties at the back and was intended to be worn under the Mae West life vest, body armor and the parachute harness. This, however, proved to be too cumbersome when worn with all the required equipment so a modified musette bag that attached to the parachute harness upon bailout was developed. After landing, the crew member could then remove and wear the vest.

C-1 EMERGENCY SUSTENANCE VEST

Three types of C-1 vests are in circulation with the two earliest models having two front slant pockets secured by a large button. The later type has these pockets opening from the top. All C-1 vests use snaps to secure the pocket flaps except for the slant-pocket design's top pockets. The earliest slant-pocket C-1 vest did not have a pocket for the insect repellent.

All components are carried in one of 14 specifically labeled and numbered outside and 2 inside pockets (items on the components list marked * are carried in the inside pockets). The first aid, fishing/sewing and accessory kits contain their specific items in tape sealed two piece clear plastic containers measuring 1" x 3" x 4" for the first aid and accessory kits and 1" x 3" x 3" for the fishing/sewing kit. A leather holster with one securing strap is located on the left side of the vest for the standard .45 caliber automatic pistol.

The only item not included in the component list is a collapsible asbestos cooking utensil that was listed on one of the inside pockets, but this item was generally never issued.

Several components of the C-1 vest were also listed for use in other seat type survival kits, such as the first aid kit, .45 caliber shot ammo, fishing/sewing kit, flares, water bladder, sharpening stone, folding sun goggles, parachute rations, mirror and a 21W survival manual.

C-1 vest (slant pocket variation)

LABEL

VEST, EMERGENCY SUSTENANCE
TYPE C-1
SPECIFICATION NO. 3206
ORDER NO. W33-038AC-5413
BRESLEE MFG. CO.
PROPERTY
"AIR FORCES U.S. ARMY"

C-1 EMERGENCY SUSTENANCE VEST

Components for the C-1 vest:

- Box of 20 .45 caliber shot ammo
- Waterproof match case with compass
- Accessory kit containing:
  - Fire starter tabs (20)
  - Pocket knife, 4" blade scout type
  - Oiler, 3 oz.
  - Plastic whistle, 2 tone
  - Plastic collapsible razor
  - Razor blades (10) (A burning lens was sometimes used in place of the razor and blades.)
- First aid kit containing:
  - Bandaids (4)
  - Two-inch compress (2)
  - Morphine syrette
  - Tube of burn ointment
  - Atabrine tablets (1 vial)
  - Salt tablets (1 vial)
  - Halazone tablets (1 vial)
  - Wound tablets (1 vial)
  - Benzadrine tablets (1 vial)
  - Instructions

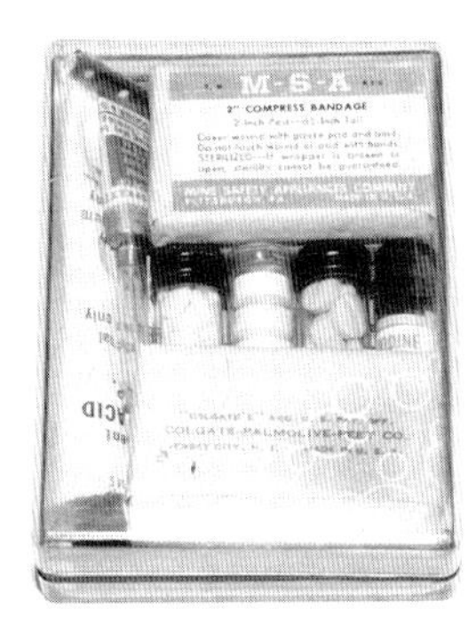

First aid kit (Enlarged from photo on page 15)

C-1 vest components, inner pockets (left to right): leather gloves with wool inserts, toilet paper in envelope, plastic pistol holster/cover, survival manual- 21W, reversible colored hat, C-1 vest instruction manual, H-1 sun goggles, 3 pint water bladder , mosquito headnet

C-1 EMERGENCY SUSTENANCE VEST

Components for the C-1 vest: (continued)

-Fishing/sewing kit
-Two bladed knife (5" cutting blade, 5" saw blade)
-ESM-2 signal mirror
-Emergency Parachute Ration (2) (5 9/16" x 3 5/16" x 1 5/16")
-Can of insect repellent
-Sharpening stone (4" x 3/8" x 3/8")
-Five minute signal flares, Fusee (2)
-Waterproof plastic pistol cover
* -Type D-3 leather gloves with wool inserts
* -Reversible color hat (yellow on one side, o.d. on the other)
* -Survival manual, 21W
* -Mosquito headnet
* -Spit assembly (This was carried in an outside pocket on some vests.)
* -Gaff assembly
* -Type H-1 folding sun goggles (also called Type II M-1943.)
* -Plastic three pint water bladder
* -Gauze bandage packet with a packet of sulfanilamide
* -Instruction/contents booklet

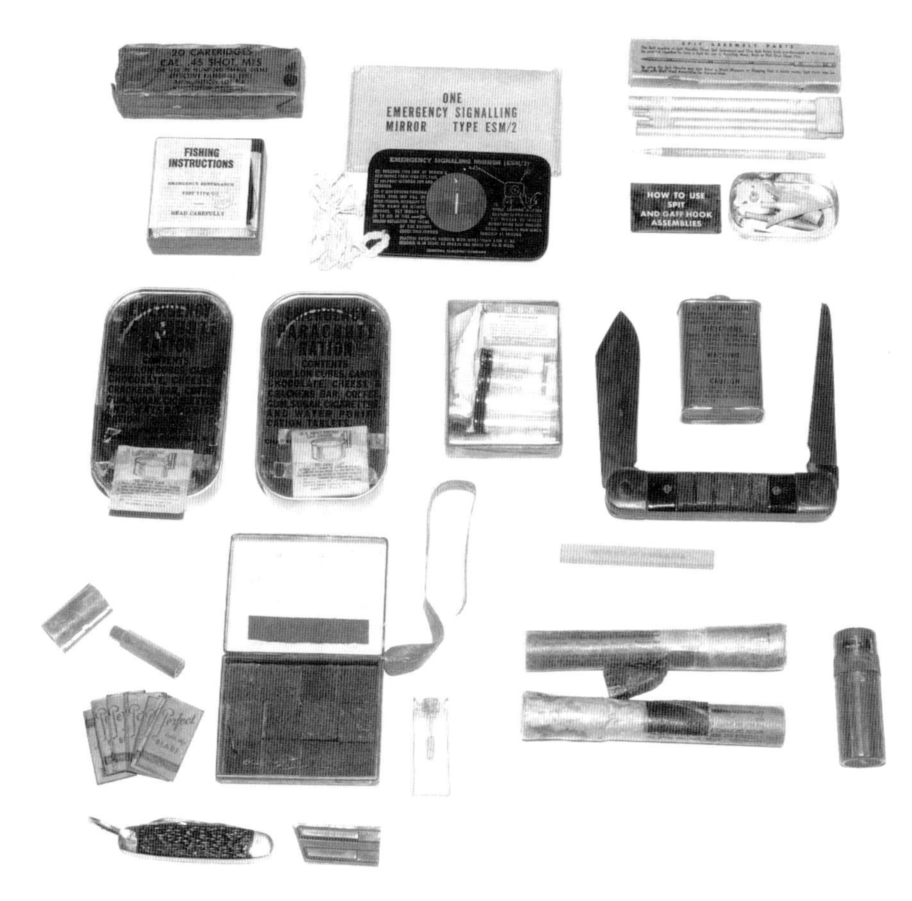

C-1 vest components, outer pockets (top row, left to right): .45 caliber shot ammo, fishing kit, ESM-2 mirror, spit and gaff assemblies (middle row, left to right): Emergency Parachute Ration (2), first aid kit, knife/saw, insect repellent, sharpening stone (bottom row, left to right): accessory kit with razor and blades, pocket knife, fire starter tabs (in bottom of accessory kit), whistle, oiler, 5 min. Fusee flares (2), compass/match case

E-1 RADIO CARRIER VEST

LABEL

VEST, RADIO CARRIER
TYPE E-1 SPEC.MIL-V-5876A
STOCK NO. 2010-978500
CONTRACT NO.AF36(600)577
BLUE ANCHOR OVERALL CO. INC.
U.S.PROPERTY EXTRA LARGE

Left: E-1 vest with URC- 4 radio and separate battery.

The U.S.A.F. E-1 Radio Carrier Vest is constructed of nylon dobby cloth and was issued in four sizes; small, medium, large and extra large. It has a zipper front closure and a nylon strap with five snaps for waist adjustment. A pocket (each with three snaps on a securing flap) located under each arm is made to hold the AN/URC-4 survival radio and it's separate battery. Three different colored vests are in circulation (dark blue, sage green and olive green).

The E-1 vest was used by aircrews in Korea and was still being listed in Air Force T.O.'s for use in multi-man raft survival kits in the late 1970's. This vest can be found in modified form with the addition of numerous pockets for survival equipment.

LABEL

VEST,RADIO CARRIER, TYPE E-1
SPECIFICATION MIL-V-5876A
SIZE MEDIUM
STOCK NO. 2010-978520
CONTRACT AF 33(600)-18057
TRU-FIT TROUSERS
U.S. PROPERTY

Right: E-1 vest modified with pockets and holster. Pocket arrangement duplicates a C-1 vest with the addition of a pocket on the left side to accommodate a larger knife.

SL-1 SURVIVAL LEGGINGS

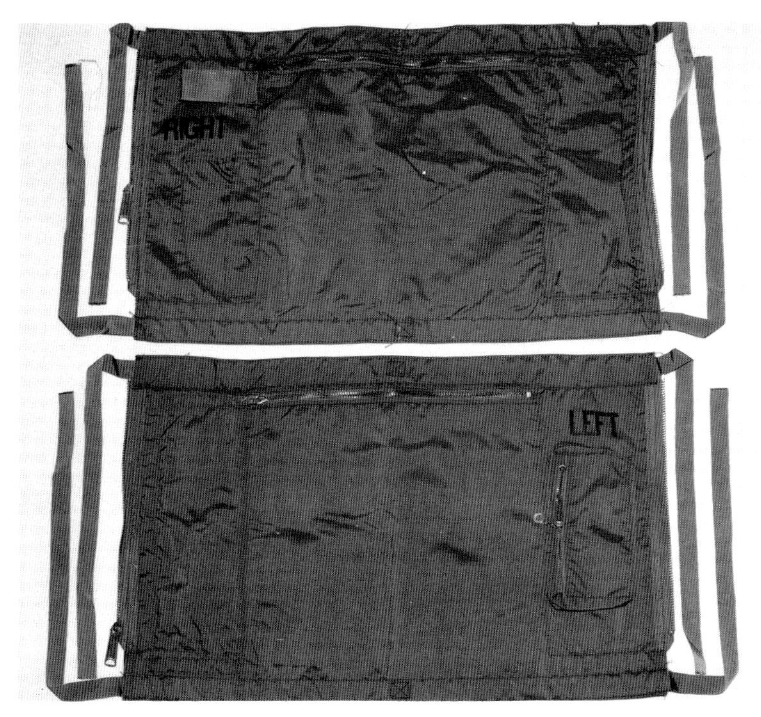

SL-1 leggings (top): right leg (bottom): left leg

LABEL

BUWEPS-U.S.NAVY
LEGGINGS,SURVIVAL,TYPE SL-1
STOCK NUMBER 8415-573-5782
CONTRACT NUMBER 5118
DSA, DIRECTORATE FOR MFG.

The U.S.N. SL-1 Survival Leggings are a 1950's design and consist of two nylon containers, a right and left, that wrapped around each lower leg. Each container measured approximately 13 1/2" x 21 1/2" opened, has a lace adjustment on back, top and bottom along with a zipper closure.

The right container has an external pocket for the pilots knife and an inner compartment (zipper closure across the top) containing three smaller pockets and one large compartment for survival gear. The left side has the same inner pocket arrangement and one external zippered pocket measuring 6 1/2" x 3".

It appears that the SL-1 was developed at the same time frame as the SV-1 vest and was designed to carried similar components because of its identical interior pocket layout.

SV-1 SURVIVAL KIT VEST

SV-1 vest, tie back (1966)

LABEL

VEST,SURVIVAL KIT,TYPE SV-1
Specification MIL-V-17893A(Wep)
Contract No. DSA IW-67-C-0479
LAN-IER INDUSTRIES, INC.
October 1966
BUWEPS-U.S.NAVY

The U.S. Navy SV-1 Survival Kit Vest was used from the late 1950's through the 1960's. It is made of sage green nylon and consists of two separate compartments, a left and right side. The vest was issued in one size and is adjustable by three ties at the back with a zipper for front closure. Some vests have elastic bands at the back instead of the adjusting ties which expanded to accommodate the varying size personnel. Suspenders supported the vest from the shoulders. Each main compartment is zippered across the top and contain three smaller compartments for survival gear. Other shades of green SV-1 vests have been found and these vests were modified with other pockets to suit individual and mission needs.

As with the SL-1 leggings (and some other vests and kits) limited information is available on the specific survival components carried. However, based upon the survival equipment available at the time, the following components would almost certainly have been used:

-SEEK 1 kit (earlier vests) or
SEEK 2 Medical and General survival kits
-L-1 or MC-1 compass
-Mk-79 type flare launcher and flares
-Mk-13 Mod 0 smoke/flares
-Whistle
-Shroud knife or MC-1 knife
-Strobe light or signal light
-Signal mirror (3" x 5" in earlier vests, 2" x 3" in later ones)
-Radio (AN/PRC-63, ACR/RT-60)
-U.S.N. tablet rations
-Pamphlet - 'Survival & Emergency Uses of Parachute'

Specific pockets for the radio, knife, etc. were not part of the basic SV-1 design but could be added.

SV-1 SURVIVAL KIT VEST

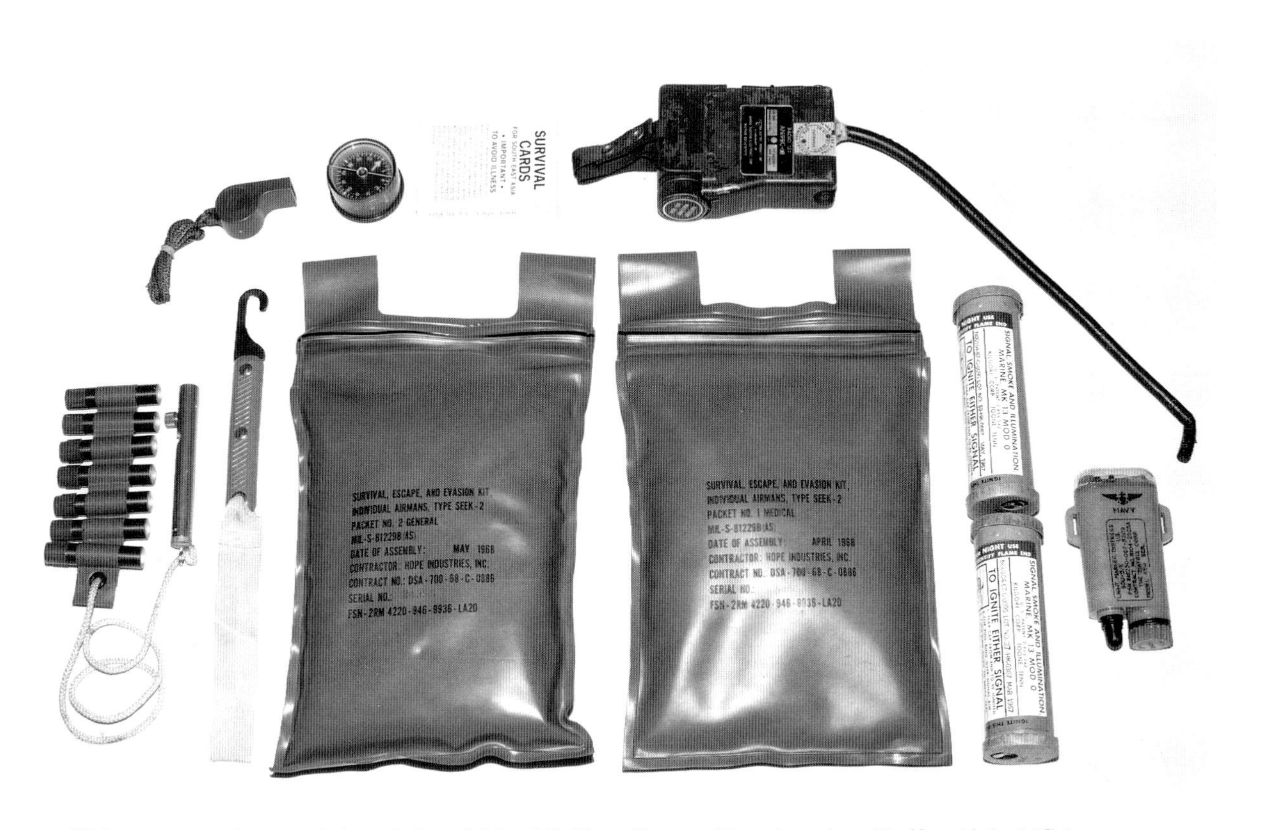

SV-1 components, later variation (left to right): Mk-79 penflares and launcher, shroud knife, whistle, MC-1 compass, S.E.A. survival cards, SEEK 2 General and Medical kits, PRC-63 radio, Mk-13 Mod 0 smoke/flares (2), U.S.N. marked strobe light

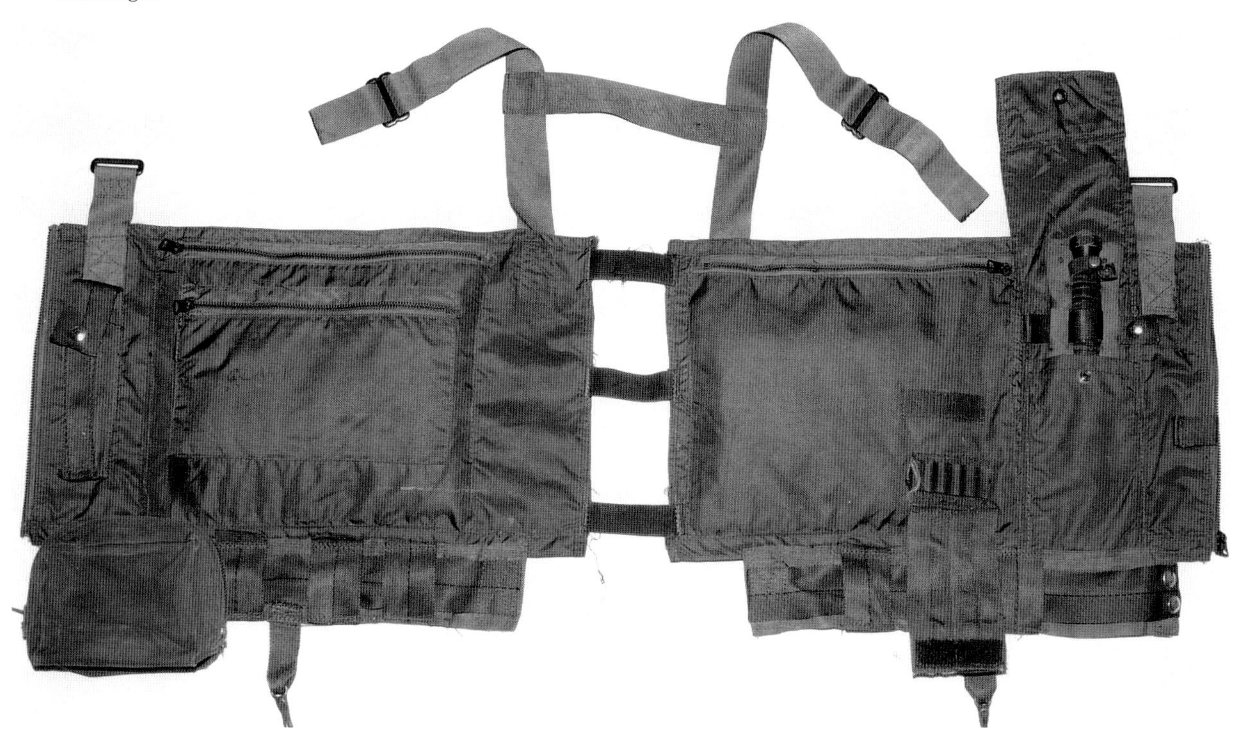

SV-1 vest modified with extra pockets and elastic bands at back

SV-2 SURVIVAL VEST

SV-2B vest with lower crotch straps modified/shortened for use in ejection seat aircraft (1987)

LABEL

VEST, SURVIVAL EQUIPMENT, TYPE SV2B
30003/67A100D2-31
SILENT PARTNER, INC.

DLA100-87-C-0338

6-87

The U.S. Navy SV-2 series survival vest, used since Vietnam, improved upon the SV-1 vest design and is still in use today by the Navy, Marine Corp and Coast Guard.

The SV-2, 2A and 2B have seen several configurations, especially in the type of mounting provided for a chest mounted oxygen regulator. The current configuration SV-2B has a zippered bag for the oxygen regulator that attaches to the front of the vest by Velcro.

This sage green nylon vest comes in one size and is fitted to an individual by modifying elastic straps on the back. The vest has attachment points for the unique U.S.N. LPA-1 through LPU-23/P 'horse collar' life preservers. The Mk-3C life preserver, which only wrapped around the waist, was used earlier. The Navy's MA-2 torso harness can be worn under the SV-2 and LP. The harness provides a rescue 'D' ring and Koch fittings for attachment to the ejection seat parachute risers. When the MA-2 harness is not used (i.e. helicopter crews, non-ejection seat aircraft, etc.), the SV-2 may be modified to include the pickup 'D' ring by the addition of a wide nylon 'belt' sewn to the upper portion of the vest, giving support for the 'D' ring.

SV-2 SURVIVAL VEST

The MA-2 harness has also been modified to incorporate several pockets containing survival equipment, thus eliminating the need for the SV-2 vest. However, the number of pockets are limited and fewer survival items can be carried. The revolver and pilots knife, if used, must be carried separately. Recently, the Navy has changed the designation of the MA-2 harness to the PCU-33.

The SV-2 components have specific compartments within the vest. The majority of items are secured within one or two larger zippered compartments with smaller dedicated pockets for each item. A nylon holster is sewn in place in a zippered pocket behind a compartment containing a 5 inch pilots knife in a leather sheath and the Mk-79 flare launcher and flares. A Velcro flap secures the knife, flare launcher and flares. Ammunition for the .38 caliber revolver is usually carried in one or two loose 9-round bandoleers tied by a nylon cord into the major compartments. The radio is placed in a fixed compartment on the left side along with a shroud cutting knife in its own pocket. The radio pocket is closed with a snap button and Velcro and after the mid- 1970's SV-2B radio pockets have a side zipper closure for easier removal and insertion of the radio. Early SV-2's had a separate battery pocket for radios using external batteries.

All items in the vest are secured to the pockets with type III nylon cord. Certain components, the shroud cutting knife and strobe light in particular, also have a five to six inch length of one inch wide type III yellow nylon cord attached to them for easier/quicker removal from their pockets.

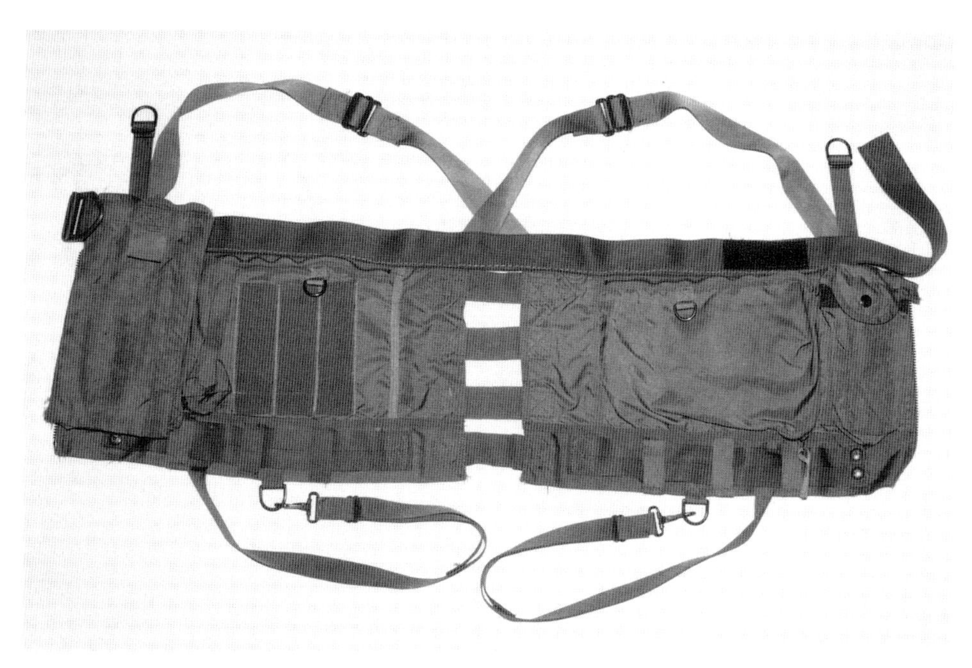

SV-2A vest (unmodified crotch straps)

LABEL

8415-144-5823
VEST,SURVIVAL, NYLON,TYPE SV2A
MIL-V-81523A, W/AMEND#1
DEFENSE PERSONNEL SUPPORT CENTER
CONT. NO. 7146-73
DATE OF MFG. FEBRUARY 1973

SV-2 SURVIVAL VEST

Components for the SV-2 series vest:

- -Radio (PRC-112, PRC-90, URC-68, PRC-63, RT-60)
- -Pilot's knife
- -Mk-79 flare launcher and 7 red flares (U.S.N. marked)
  (gold colored pengun flare launcher carried earlier)
- -.38 caliber revolver and ammunition
- -SDU-5/E strobe light
- -Mk-3 Type 1, 2" x 3" signal mirror
- -Whistle
- -Mk-13 Mod 0 smoke/flare (Also carried in flare pouch of LP.)
- -Four oz. plastic water bottles (1 or 2)
- -Two part Medical and General survival kits (SEEK 2 or SRU-31/P Airman's
  24 hr. kit)

Additional items may be added to the vest but could not exceed five pounds in added weight.

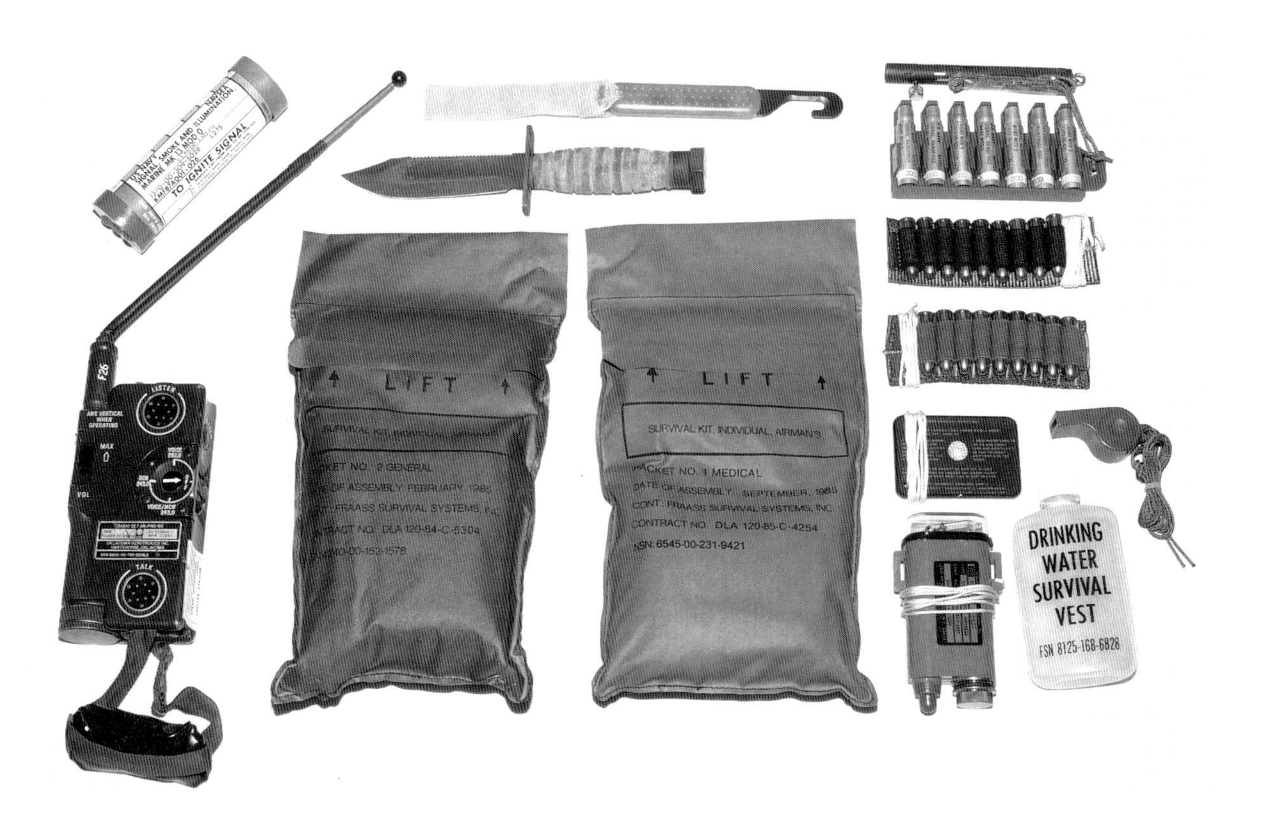

SV-2B vest typical components (left to right): PRC-90 radio, Mk-13 Mod 0 smoke/flare, shroud knife, pilots knife, SRU-31/P Medical and General survival kits, Mk-79 penflares and launcher, ammunition in bandoleers (2), 2" x 3" mirror, whistle, strobe light, 4 oz. water bottle

SV-2 SURVIVAL VEST

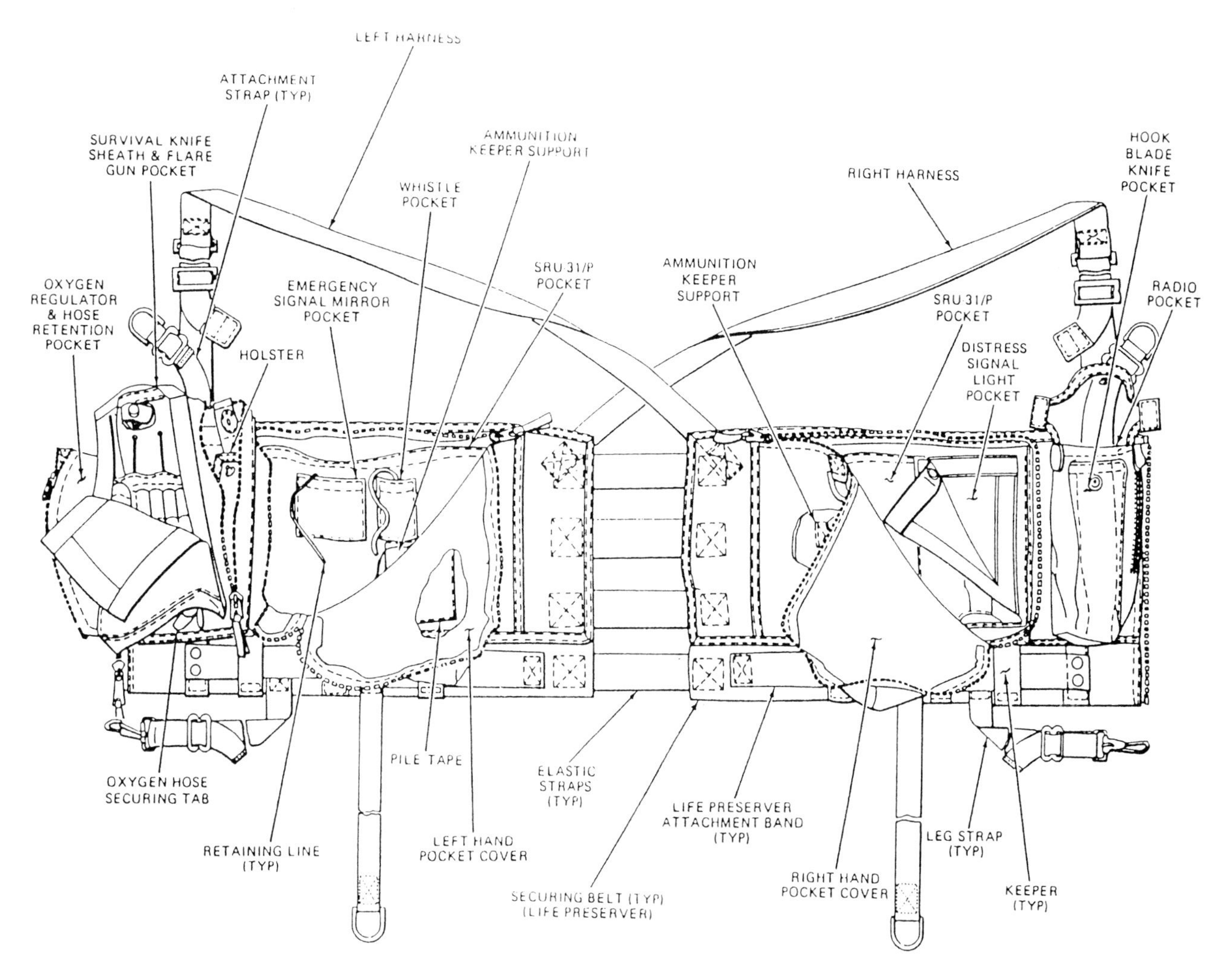

SV-2B vest parts nomenclature

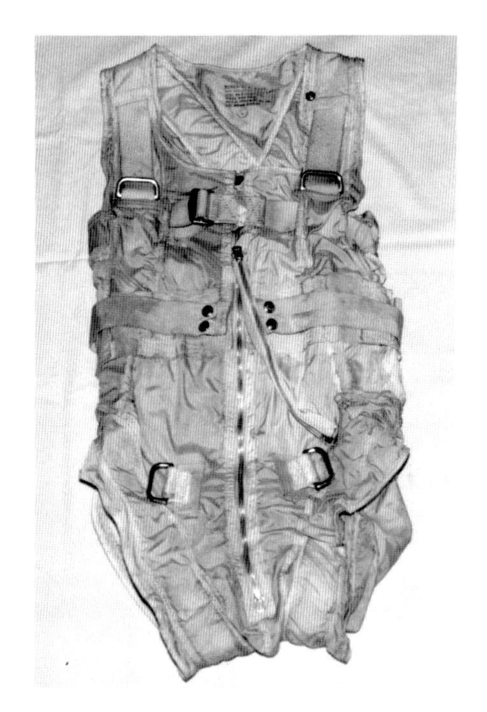

LABEL

BUAER U.S. NAVY
SUIT INTEGRATED TORSO HARNESS
SPEC. MIL-S-19089 (AER)
CONTR. N156-39303
SWITLIK PARACHUTE CO., INC.
SIZE MEDIUM REGULAR

Left: Late 1950's, early 1960's issue white nylon torso harness (Note difference in fittings compared to later Koch fittings.)

SV-2 SURVIVAL VEST

Modified MA-2 torso harness typical components (left to right): PRC-90 radio, angle head flashlight, Mk-79 pen-flares and launcher, 4 oz. water bottle, Mk-13 Mod 0 smoke/flare, shroud knife, strobe light, 2" x 3" mirror, whistle

MA-2 torso harnesses (left): current issue type (1984), (right): earlier issue (1973)

LABELS

MA-2 TORSO HARNESS ASSY. SIZE:LR
DESIGN ACTIVITY CODE 30003
PART NO. 829AS142-8LR MFR. 77715
CONT. NO. N00383-83-C-1611
DATE OF MFR. MAR 1984
DATE PLACED IN SERVICE
COMP. ASSY. 829AS100

BUWEPS U.S. NAVY
SUIT, INTEGRATED TORSO HARNESS MA-2
SPECIFICATION NO. MIL-S-19089
STOCK NO. IRD1670-866-4344-LX3X
CONTRACT NO. N00383-73-1175
PIONEER RECOVERY SYSTEMS, INC.
SIZE 3 SL
DATE OF MFR. SEPT 1973

SV-2 SURVIVAL VEST

Components for the modified MA-2 torso harness:

- -Radio (usually PRC-90)
- * -Water bottle
- * -Mk-3 Type 1, 2" x 3 signal mirror
- * -Mk-13 Mod 0 smoke/flare
- * -Whistle
- * -Mk-79 flare launcher and flares
- -Angle head flashlight with red lens
- -SDU-5/E strobe light
- -Shroud knife

Items marked with * are carried in one compartment on the left. All other items have their own specific pocket.

Other items that could be carried in the SV-2 vest and the MA-2 modified harness either in addition to or in lieu of standard items:

- -MC-1 compass
- -Dyemarker
- -Small chemical lights
- -MC-1 orange handle switch blade knife
- -Pen light
- -Space blanket

The Medical and General kits could be eliminated from the SV-2 vest; and the dye markers and Mk-13 Mod 0 smoke/flares may be carried in pouches attached to the LP's. The plastic 4 oz. water bottles are being replaced by 4.2 oz. sealed foil containers.

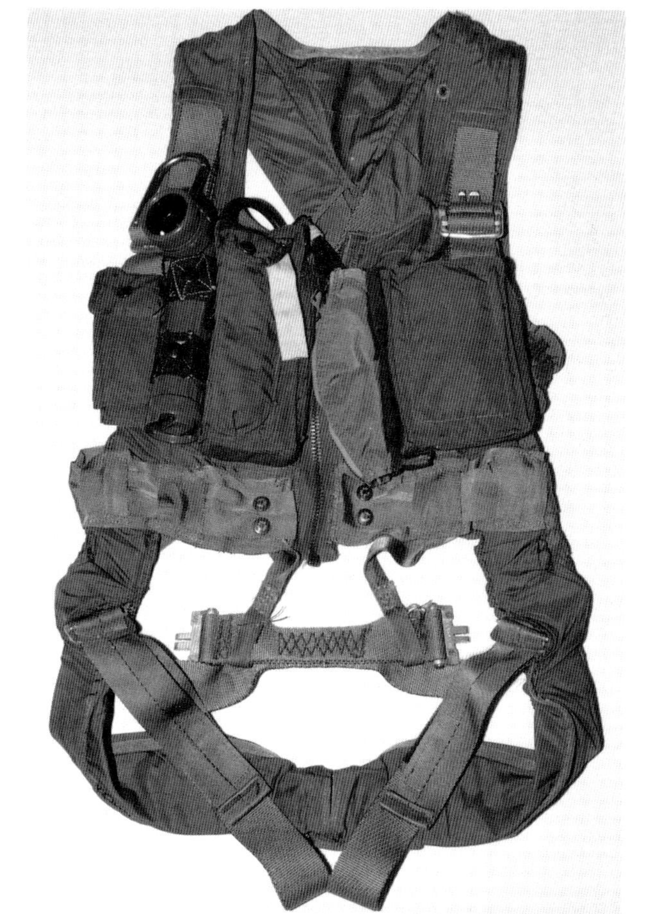

Right: Modified MA-2 torso harness

CMU-24/P SURVIVAL VEST

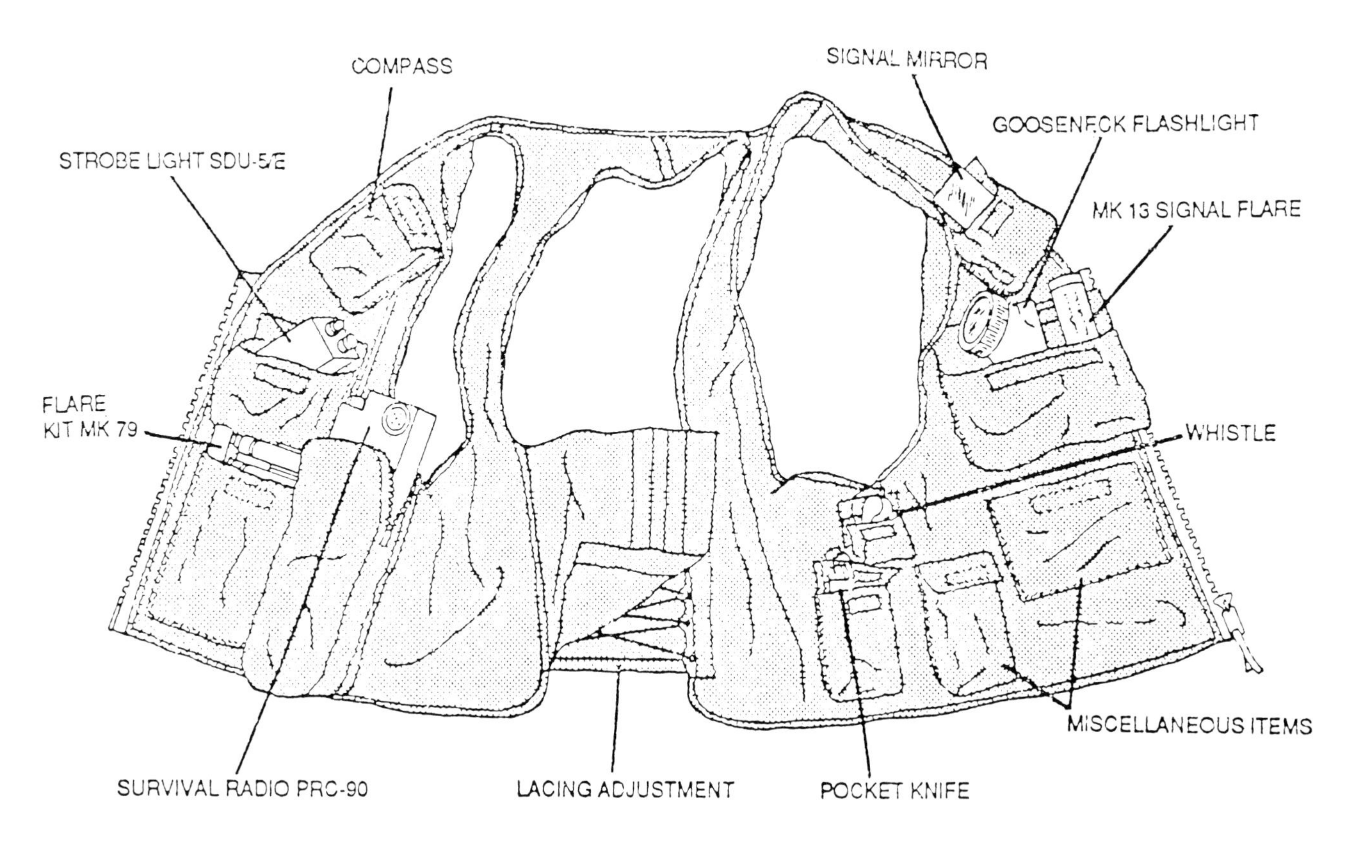

Location of survival items in CMU-24/P vest

The U.S. Navy CMU-24/P Survival Vest is a modified SRU-21/P and is designed for use by its search and rescue aircrew members.

Components are:

- -Mk-3 Type 1, 2" x 3" signal mirror
- -Mk-13 Mod 0 or Mk-112 smoke/flare
- -Whistle
- -Pocket knife
- -Lensatic compass
- -SDU-5/E strobe light
- -Mk-79 flare launcher and flares
- -Radio (PRC-112 or PRC-90)
- -Flashlight

As with the SV-2, additional items can be added to the vest by the crew member, but cannot exceed five pounds. All items are secured to the vest by specific lengths of type I nylon cord. The strobe light has an addtional piece of 5 to 6 inch length of type II yellow or type IV sage green one inch wide nylon attached for easier handling.

SRU-19/P SURVIVAL CHAPS

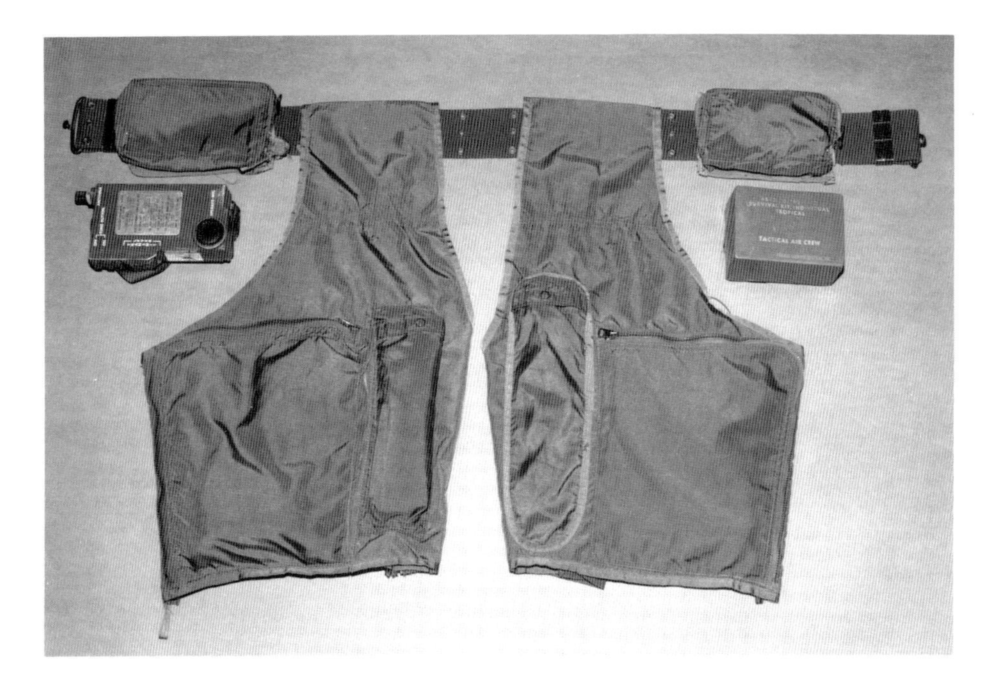

SRU-19/P chaps showing arrangement of chaps, radio pouch (RT-10) and 'TAC' kit pouch on web belt

The Air Force SRU-19/P Survival Chaps were procured in medium and large sizes and consisted of two sections, a left and right, which were worn around the thigh of each leg. The chaps, constructed of green nylon duck, were channeled through the standard cotton webbed belt for suspension and adjusted by lacing. Zippers affected the closure around each leg.

The radio and the Individual Tropical Survival Kit (TAC kit) with signal mirror were carried in two separate zippered pouches. The web belt used for suspension of the chaps was also routed through each pouch. The .38 caliber revolver was carried in a leather holster attached with Velcro to the inside of the left chap. The five inch blade knife, water bladder and gill net were carried in their own pockets. All other components were positioned in specific snap closure pockets within the zippered main compartments.

The chap containers could be converted into a vest container for carrying purposes upon reaching the ground after bailout. The SRU-19/P chaps were used during Vietnam by aircrews of A1E, O1E/F, B-52, KC-135, U-10B, T-38, RF/F4, F-100, F-105, RB/B-57, RB-66 and RF-101 aircraft.

SRU-19/P SURVIVAL CHAPS

Components for the SRU-19/P:

- -URC-64 radio or ARC/RT-10 radio as substitute
- -Acoustical coupler (RT-10 radio only)
- -Earphone w/carrying case (URC-64 radio only)
- -SDU-5/E light marker w/flashguard FG 1B
- -Signal kit, type A/P22S-1 (1370-921-6172-LY35)
- -Individual Tropical Survival Kit (TAC kit)
- -Mk-3 Type 1, 2" x 3" signal mirror
- -Tourniquet (non-pneumatic type)
- -Lensatic compass
- -Gill net
- -Pocket knife
- -Three pint plastic water bladder, size B
- -Fire starter, type M-2 (The Zippo type lighter or match container w/matches could be used in lieu of the M-2 fire starters.)
- -Insect repellent, camouflage stick
- -Five inch blade pilots knife
- -.38 caliber revolver w/ammunition (23 rounds including 6 rounds of tracer)

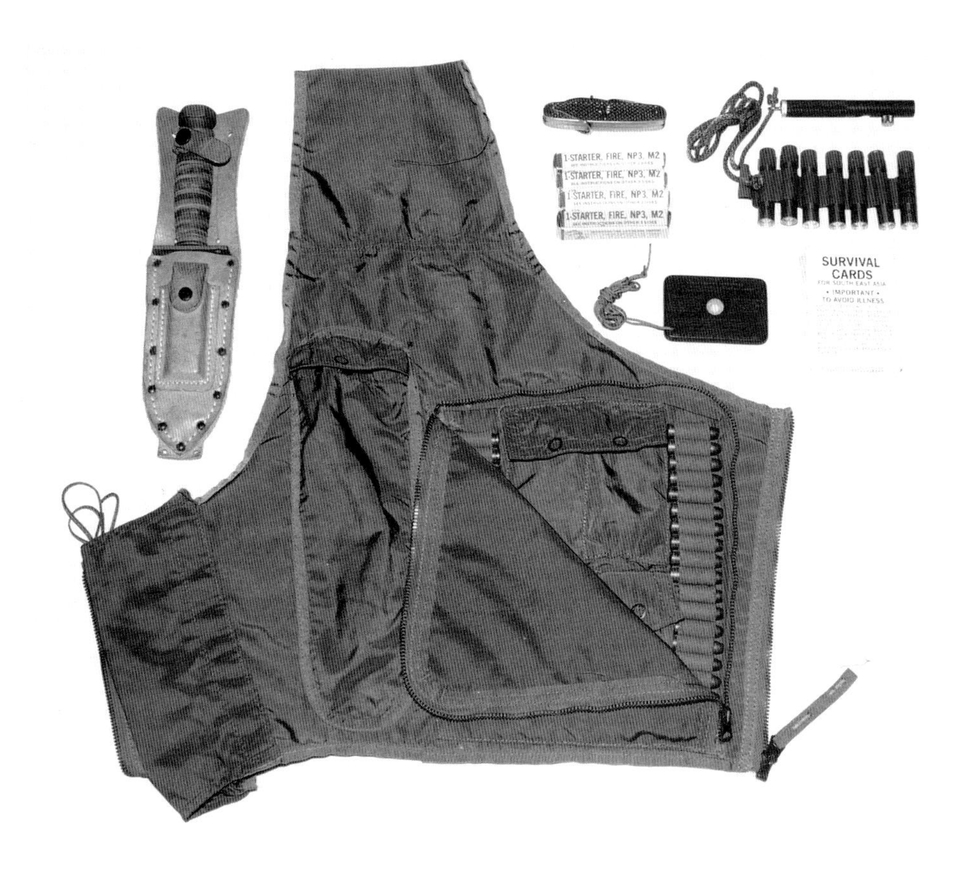

LABEL

CHAPS, SURVIVAL
SIZE: LARGE
IRVING AIR CHUTE CO., INC.
AF 36(657)15402
RIGHT CHAP

SRU-19/P right chap (left to right): pilot's knife in sheath, pocket knife, M-2 fire starters (4), 2" x 3" mirror (can be carried in TAC pouch), Southeast Asia (S.E.A.) survival cards, penflares and launcher (Ammunition is placed in the keepers in the chap.)

SRU-19/P SURVIVAL CHAPS

SRU-19/P left chap (left to right): tourniquet, lensatic compass, match case, strobe and strobe cover, leather holster, 3 pint water bladder, gill net in bag

LABEL

CHAPS, SURVIVAL
SIZE: LARGE
IRVING AIR CHUTE CO., INC.
AF 33(657)15402
LEFT CHAP

SRU-20/P SURVIVAL VEST

SRU-20/P vest (1970)

LABEL

VEST, SURVIVAL
INTEGRATED HARNESS
SIZE UNIVERSAL
IRVING AIR CHUTE CO., INC.
AF 33(657)15402

Used during Vietnam, the U.S. Air Force SRU-20/P Survival Vest was compatible only with the integrated harness of RF/F-4 aircraft and was used in lieu of the SRU-19/P chaps.

The vest was fabricated of sage green nylon duck material and consisted of two sections, a left and right panel. The top portions of both sections channel through the harness of the underarm life preserver LPU-2/P or LPU-10/P and the vest came in only one size. The girth was adjusted by lacing. The various pockets containing the survival equipment were secured by either snaps or zippers.

Components for the SRU-20/P:

Right Side

-Radio (RT10, URC-64, etc.)
-Pilots knife
-Fire starter
-Ammunition
-Distress signal/strobe light - SDU 5/E & flashguard FG 1B

Left Side

-Tourniquet (non-pneumatic type)
-Lensatic compass
-Three pint water bladder, size B
-Gill net
-Holster and 20 rounds of ammmunition
-Flare launcher and flares (standard pengun type)
-Pocket knife
-Signal mirror, Mk-3 Type 1
-TAC kit (Individual Tropical Survival Kit)

SRU-20/P SURVIVAL VEST

SRU-20/P vest (Closure flaps are opened to show inner arrangement.)

SRU-21/P SURVIVAL VEST

SRU-21/P vest with PRC-90 radio pouch and pilots knife (1974)

SRU-21/P vest (back)

LABEL

8415-933-6231
VEST,SURVIVAL,NYLON
MESH NET, SRU-21/P
SIZE: MEDIUM
LITE INDUSTRIES, INC.
CSA 100-74-C-0033
USAF DRAWING NO. 66D1596

The SRU-21/P Survival Vest was introduced into the Air Force during the Vietnam war for use by all their aircrews. The U. S. Army also used the vest for its helicopter crews and it proved beneficial to special ground forces as well. The Air Force and Army are still using the SRU-21/P today and the Navy even uses a modified version for its search and rescure aircrews designed the CMU-24/P. The Army, however, will be replacing the vest with a new design called the SARVIP that incorporates body armor and an extraction harness.

This nylon cloth mesh, sage green vest comes in medium and large sizes with a string adjustment in back and incorporates twelve pockets (10 outside, 2 inside) for a variety of survival items. These pockets can be relocated and modified to suit the mission or the wearer. There is also an Aramid mesh fabric fire resistant vest available. The pockets use Velcro for closures but very early designs had snap pockets. A leather holster with two securing straps is normally sewn onto the left side to accommodate a .38 caliber revolver. Earlier designed holsters had only one strap. The nylon OV-1 vest holster is also being used in lieu of the leather holster. The extra ammo resides in one of the smaller pockets. The standard five inch pilots knife with leather sheath and sharpening stone is sometimes added to the vest if not carried on the individual. Both items can be sewn in a particular position to suit the airman. All components are secured to the vest by a specific length of type III nylon cord or size III cotton cord.

The vest can be worn over body armor and an underarm life preserver, type LPU-2/P or LPU-10/P. An LPU-3/P is also used but integrates into a parachute torso harness which attaches to the harness with zippers. The newest life preserver, the LPU-9/P, is worn around the neck and chest and attaches to the parachute torso harness.

SRU-21/P SURVIVAL VEST

Components for the SRU-21/P vest vary depending upon the mission or command. Because of the many variations, components below are those outlined in the basic operating manuals TM 55-8465-215-11 (dated June 1970) and TM55-1680-351-10 (dated 22 April 1987). The operating manual and a copy of the assembly sheet drawing with FSN/NSN's are sometimes placed in an inside pocket. Copies of the assembly sheet drawings are included in the appendix.

Components for the SRU-21/P:

- -Individual Tropical Survival Kit (TAC)
- -Tourniquet (non-pneumatic type)
- -Gill net
- -Three pint water bladder, size B
- -Mk-3 Type 1, 2" x 3" signal mirror
- -SDU-5/E distress marker light (strobe light) w/AGR-FG1B flash guard
- -Foliage penetrating signal kit (M-201, rocket launcher and seven red flares) Other types of pengun signal kits were used earlier.
- -Pocket knife
- -Fire starter (butane lighter, match case, M-2, magnesium fire starter, Aviation Spark-Lite)
- -.38 caliber revolver or 9mm pistol and ammunition (17 rounds ball M41, 6 rounds tracer M41)
- -Radio (The current radio is the PRC-112. The AN-PRC-90, AN-URC-68, ARC-RT-10, and URC-64 have also been used.)
- -Whistle w/lanyard

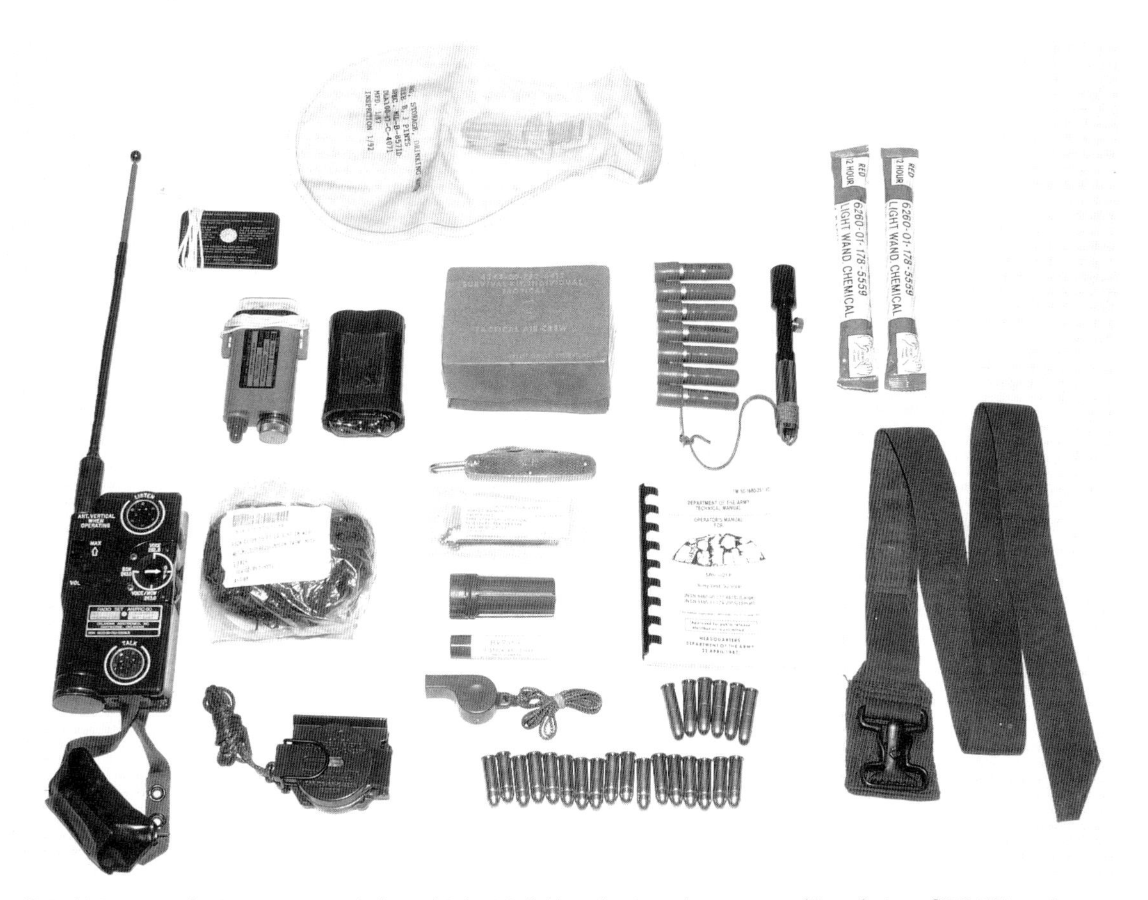

SRU-21/P vest typical components (left to right): PRC-90 radio, lensatic compass, gill net in bag, SDU-5/E strobe light, strobe cover, 2" x 3" mirror, 3 pint water bladder, Individual Tropical Survival Kit, pocket knife, magnesium fire starter, match case, chapstick, whistle, ball ammo (17 rounds), tracer ammo (6 rounds), operators manual, 'rocket' type penflare and launcher, chemical light wands (2), tourniquet

SRU-21/P SURVIVAL VEST

Other survival components that can sometimes be found in the SRU-21/P are:

- -Chapstick
- -Insect repellent
- -Emergency blanket (combat casualty blanket)
- -Insect sting kits
- -Camouflage sticks
- -Snake bite kits
- -Pen lights
- -Chemical light sticks
- -Mk-13 Mod 0 smoke/flares
- -Pilot's knife
- -MC-1 compass (in lieu of lensatic)

Label

VEST,SURVIVAL
DLA 100-85-C-05466
100% ARAMID
8415-00-177-4818
LITE INDUSTRIES, INC.

Right: SRU-21/P vest of Aramid fabric (1985)

LABEL

8415-00-933-6232
VEST,SURVIVAL,NYLON,MESH
NET,SRU 21/P
LARGE
LANCER-CLINTON CORP.
DLA100-79-C-2936
USAF DRAWING NO. 66D1596, REV. 1

Left: SRU-21/P vest with RT-10 type radio pouch (1979)

SRU-21/P SURVIVAL VEST

SRU-21/P vest of early snap pocket design

LABEL

VEST,SURVIVAL, MESH NET, SRU-21/P
8415-00-1774818
LARGE
AVIREX, LTD
DLA 100-83-C-0708
USAF DRAWING NO. 66D1596 REV 1

Left: SRU-21/P vest modified as a pickup harness (1983).

SRU-32/P FRAGMENTATION SURVIVAL VEST

SRU-32/P vest (1970)

LABEL

VEST,FRAGMENTATION - SURVIVAL
SRU 32/P
CONTRACT F33657-70-C-0780
SIZE: EXTRA LARGE
LANCER CLOTHING CORP.

The SRU-32/P Fragmentation Survival Vest was developed in the late 1960's to provide frontal protection against small arms fire and shrapnel plus contain any needed survival gear required for the mission. The sage green vest contains twelve pockets (10 outside, 2 inside) similar in design to the SRU-21/P and had a mesh back.

The vest was donned by passing the head through an opening that separates the front and back of the vest and once in place, straps from the back wrapped around the front and were secured by Velcro. Velcro on the radio and TAC kit pockets aided in holding the strap. The vest contained the same survival components as the SRU-21/P; however, the radio pocket was designed for the RT-10 radio or equal.

FIGHTER ATTACK VEST

Fighter Attack Vest (1970)

LABEL

TWC FIGHTER/ATTACK
FRAGEMENTATION VEST
CONTRACT: F33657-70-C-0713
SIZE: SMALL REGULAR
NO FSN

The U.S.A.F. Fighter Attack Vest is similar in design to the SRU-32/P vest but is lighter in weight and easier to don and remove. The individual first slips the right arm through the arm opening on the right side of the vest, then wraps the mesh fabric back portion around to the left. The back section attaches with Velcro at the left upper shoulder, and Velcro and a snap on the left side. This system allows for quick removal of the vest. A 'Spandex' material on the back panel gives greater flexibility in movement.

The very thin and flexible armor plating is incorporated into the front of the vest and the survival pockets are arranged like the SRU-32/P. This armor plating is removalable and is held in place by a Velcro flap at the bottom front of the vest. The RT-10 radio was used and the other survival items are the same as the SRU-21/P and 32/P.

ARMY LIGHTWEIGHT INDIVIDUAL SURVIVAL KIT

'Leg holster kit' nylon container/carrying bag with attached plastic coated components list card (reverse side of card also displayed - showing medical information), nylon carry bag for Auxiliary Survival Kit (upper right)

LABEL

HOLSTER
SURVIVAL KIT, INDIVIDUAL
LIGHTWEIGHT,ARMY
11-1-943
ROCKET JET ENG. CORP.
F.S.N.8465-J01-0741

LABEL

BAG, CARRYING
SURVIVAL KIT, INDIVIDUAL
LIGHTWEIGHT,ARMY
11-1-946
ROCKET JET ENG. CORP.
US

The Army Lightweight Individual Survival Kit (leg holster kit) was used during Vietnam by Army aircrews on helicopters and scout planes. The kit was suspended from a belt and worn around one leg and once on the ground could be slung over the shoulder for easier carrying.

The kit consisted of a three pocket nylon container/carrying bag with Velcro closures to secure the pocket flaps. An Auxillary Survival Kit and instruction/components card was placed in the largest pocket, a survival radio (usually an RT-10) fit into another smaller one and a pilot's knife and distress marker light occupied the third pocket.

ARMY LIGHTWEIGHT INDIVIDUAL SURVIVAL KIT

The Auxillary Survival Kit consisted of a trapezoidal shaped two piece container measuring 6" wide (top) x 4 1/4" wide (bottom) x 8" high x 2" thick and was curved to conform to the wearer's thigh. The two piece container was held together by a snap on each end and fit into a separate nylon carrying bag with belt loops.

Forty-four individual items were in sealed plastic bags and attached by a tacky 'glue' to a four-section folded sheet within the bottom plastic part of the container. The two-piece container fits into a one quart clear vinyl water container that had belt carrying loops and used a ziplock closure. A signal mirror, flare launcher and four flares were taped to the rear outer surface of this container. The teflon coated metal lid of the container was designed to be used as a cooking utensil with the flare launcher being attached by a threaded fitting to act as a handle. In addition, the flare launcher was used as a handle for the small hacksaw blade and knife blade supplied in the kit, attaching in a similar manner.

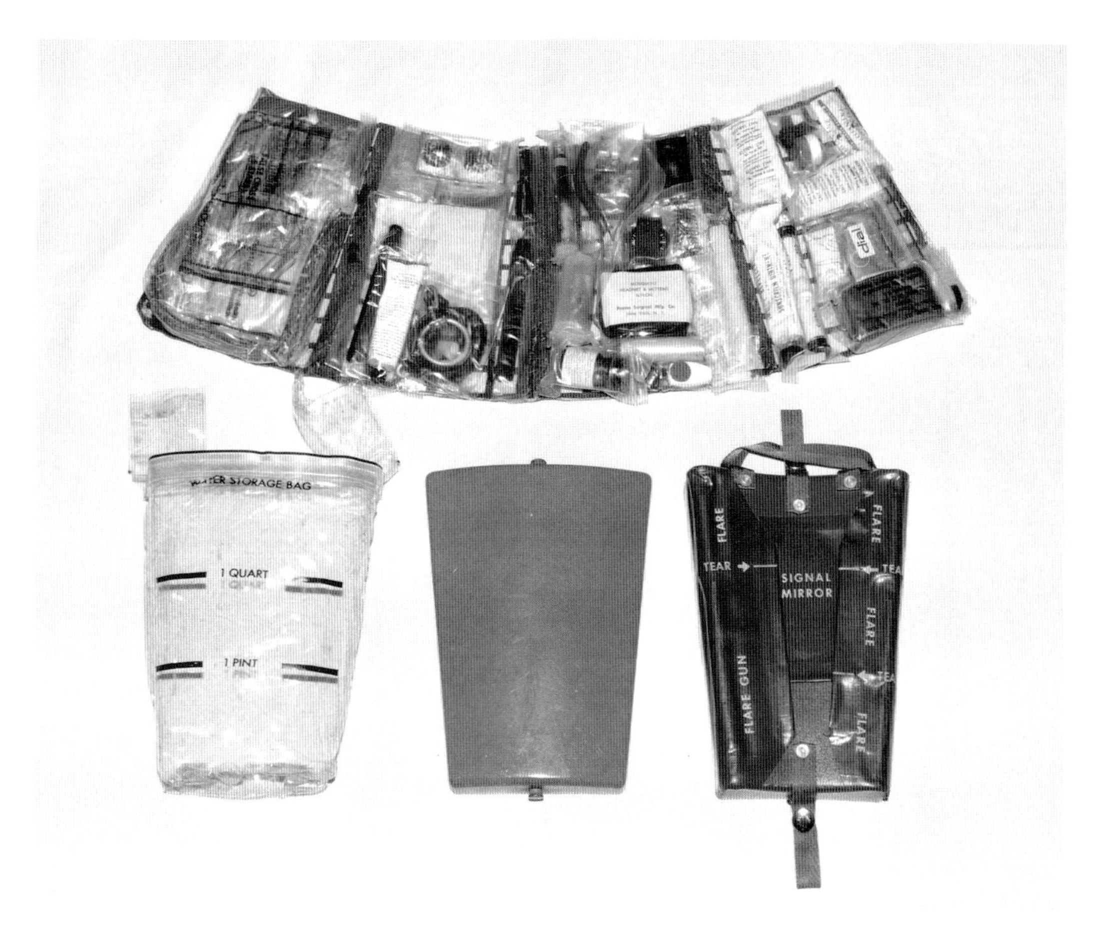

Auxilary Survival Kit (top row): components (bottom row, left to right): 1 quart water bag, teflon coated lid, back of bottom container showing placement of mirror, flare launcher and 4 flares

ARMY LIGHTWEIGHT INDIVIDUAL SURVIVAL KIT

Auxiliary Survival Kit opened to show components.

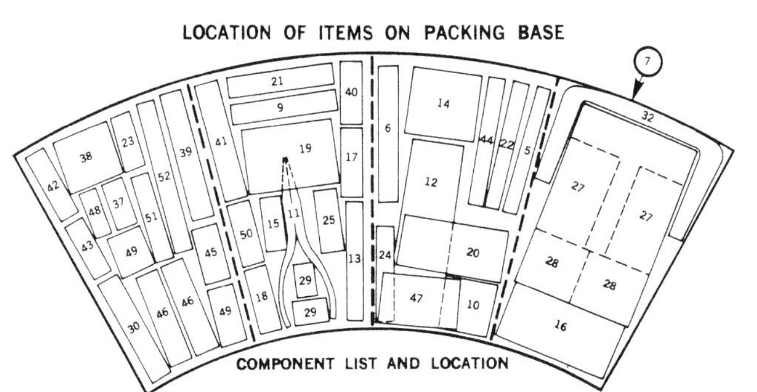

1. Lid assembly (not shown)
2. Container assembly, lower (not shown)
3. Mirror assembly, signal (in pocket on item 2)
4. Bag, water storage (over item 7)
5. Blade assembly, knife
6. Blade assembly, hacksaw
7. Base, component packing
8. Flare projector and 4 flares (in pockets on item 2)
9. Waterproof matches and synthetic flint
10. Firestarter sheets and tinder
11. Needle nose pliers
12. Aluminum foil
13. Sewing kit
14. Wire saw
15. Wrist compass and strap
16. Nylon cord
17. Fishing kit
18. Fishing line
19. Mosquito head net and mittens (compressed)
20. Compressed sponge
21. Flashlight and lanyard
22. Candle
23. Toothbrush and dentifrice
24. Snare wire
25. Plastic whistle and lanyard
26. Instruction booklet and pencil (in pocket on item 34)
27. Salt and pepper
28. Chili powder
29. Bouillon cubes (4)
30. Waterproof receptacles
31. Solar still
32. Plastic tubing (4 ft.)
33. Bag, carrying (not shown)
34. Survival, holster (not shown)
35. Knife, hunting, sheathed (in pocket on item 34)
36. Card — instructions and components (in pocket on item 34)
37. Antiseptic soap
38. Elastic gauze bandage (2″ by 5 yd.)
39. Sun and bug repellent (1 oz. tube)
40. Water purification tablets (50 tablets)
41. Adhesive bandages (6)
42. Adhesive tape (1″ by 1 yd.)
43. Eye ointment (1/8 oz. tube)
44. Antibiotic ointment (1/2 oz. tube)
45. Anti-malaria tablets (3)
46. Pain killer capsules (12)
47. Anti-diarrhea tablets (12)
48. Stay awake tablets (10)
49. Salt tablets (20)
50. Anti-infection tablets (12)
51. Tweezers and safety pins
52. Reflective chalk (2)
53. Light, marker, distress (in pocket on item 34)

OVER

Left: Plastic coated components list card (reduced)

ARMY LIGHTWEIGHT INDIVIDUAL SURVIVAL KIT

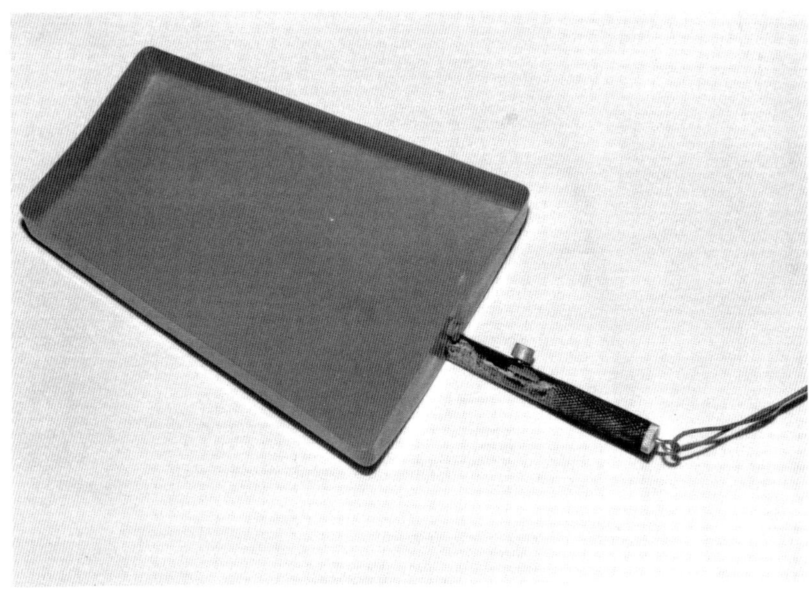

Teflon coated lid of Auxiliary Survival Kit shows attachment of flare launcher for use as a handle.

AUXILIARY SURVIVAL KIT LABEL

SURVIVAL KIT, INDIVIDUAL
LIGHTWEIGHT, ARMY

ROCKET JET ENGR. CORP.
PART NO. 776000-1
PAT.PEND.
CONTRACT NO. DAAG.17-67-C-0056

MEDICAL INFORMATION

PAIN KILLER CAPSULES: Use for headaches and acute pain. Take one capsule every three hours. No more than 6 capsules in 24 hours.

SALT TABLETS: For excessive sweating take 1 or 2 tablets with a drink of water. Use 4 to 6 times daily as needed. Swallow whole.

ANTI-INFECTION TABLETS: For fever, infection, diarrhea, take 1 tablet every 4 hours.

WATER PURIFICATION TABLETS: Not for use with salt water. Add 1 tablet for each quart of clear water, 2 tablets for each quart of water if not clear. Wait 5 minutes, shake well, wait 10 minutes before using for any purpose. For very cold water wait 20 minutes.

EYE OINTMENT: In case of inflammation or foreign body which cannot be easily removed, squeeze a small amount of ointment on inner surface of lower lid, apply 2 or 3 times a day. CAUTION: DO NOT RUB EYE.

ANTIBIOTIC OINTMENT: Apply to minor cuts, wounds, burns and abrasions once or twice daily, and cover with dressing.

ANTI-MALARIA TABLETS: Take 1 tablet weekly to prevent Malaria. Take with or after food. Swallow, do not chew.

STAY AWAKE TABLETS: To keep awake take one tablet every 12 hours.

ANTI-DIARRHEA TABLETS: For diarrhea take two tablets three times a day.

Left: Reverse of components list card shows medical information.

OV-1 AIRCRAFT SURVIVAL VEST

OV-1 vest (1973) (Life preserver is not installed.)

LABEL

SURVIVAL VEST OV-1 AIRCRAFT
PART NUMBER
MCN 8415-H06-2043
DATE OF MFR.

MFR. NAME WELSON & CO., INC.
SIZE SMALL

The OV-1 Aircraft Survival Vest is unique because it was designed to be used for a specific aircraft, namely the Army's OV-1 Mohawk and to be used in conjunction with its ejection seat and survival kit.

The vest is constructed of green Rashel-knit nylon cloth and was furnished in two sizes, small and large. The OV-1 vest resembles the Navy SV-2 series vest with later OV-1 versions having a mesh suspension arrangement added above the main body. The LPU-2/P or LPU-10/P underarm life preservers are integrated into the vest by routing the chest strap of the LPU's through channels in the upper portion of the right and left body of the vest. The vest is also worn on top of the torso harness. Adjustment is by lacing on the back and a zipper closure on the front.

A removable nylon holster for a .38 caliber revolver is placed in a zippered pocket behind a five inch pilots knife and the rocket type flare launcher with flares. The pockets for the knife and flares are Velcro closures. The radio has its own pocket on the right side and an MC-1 'switch' blade/shroud line knife is placed in a smaller pocket attached to the radio pocket. A whistle resides in a pocket under the radio pocket. A zippered right and left main compartment contain the SRU-31/P Medical and General kits, signal mirror, strobe light, and ammunition bandoleer.

Once on the ground the holster can be removed from its compartment and secured to the right outside portion of the vest by two snaps and a leg lanyard. The ammunition bandoleer can be removed from within the left main compartment and attached by snaps to the left side of the vest.

OV-1 AIRCRAFT SURVIVAL VEST

Components for the OV-1 vest:

- -SDU-5/E strobe light w/flashguard
- -Rocket type flare launcher and 7 red flares (M-201)
- -Underarm life preserver (LPU-2/P, or LPU-10/P)
- -MC-1 knife
- -Whistle
- -Mk-3 Type 1, 2" x 3" signal mirror
- -Five inch blade pilot's knife
- -Radio (PRC-90, RT-10 or URC-68)
- -.38 caliber revolver and 12 rounds of ammunition (6 ball and 6 tracer)
- -SRU-31/P survival kits (General Individual Kit and Medical Individual Kit)
- -Magnesium fire starter
- -Operators manual TM55-1680-316-10

OV-1 components (left to right): pilot's knife, ammunition bandoleer and ammo, whistle, operators manual, Airman's Individual Survival Kits (Medical and General), butane lighter and lid, match case , MC-1 switchblade/shroud knife nylon holster, RT-10 radio, 2" x 3" mirror, 'rocket' type penflare and launcher, strobe light and cover

OV-1 AIRCRAFT SURVIVAL VEST

OV-1 vest showing holster and ammunition bandoleer attachment
Life preserver is not installed.

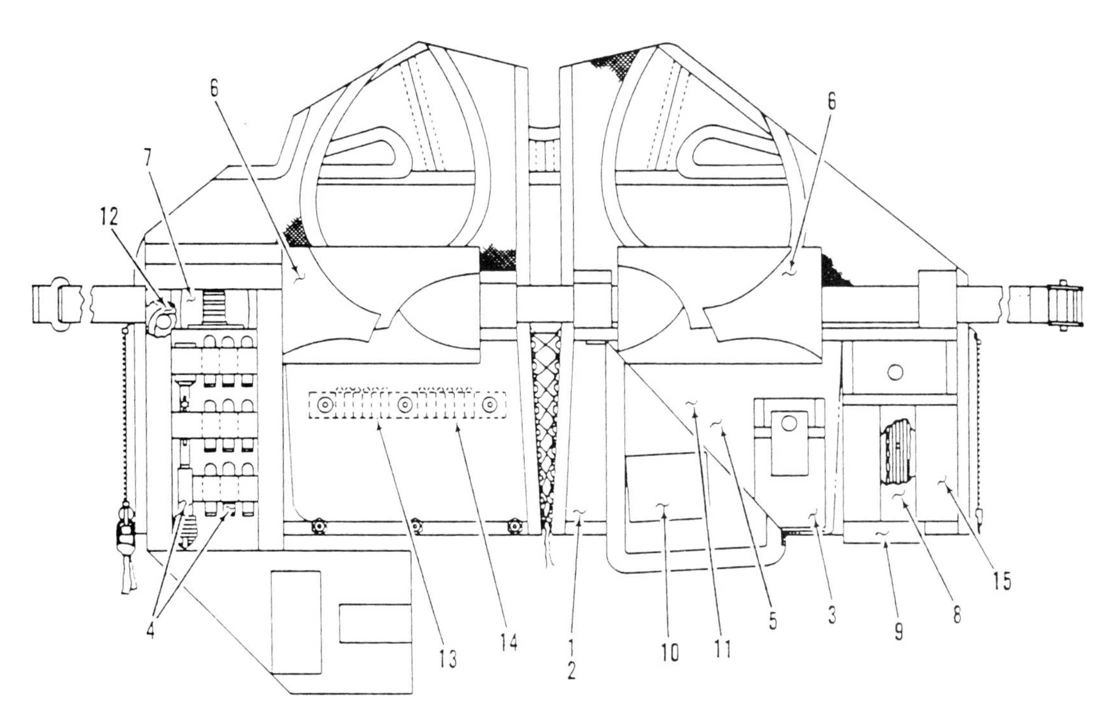

1. Survival vest, small
2. Survival vest, large
3. Distress marker light (SDU-5/E)
4. Distress signal kit (M-201)
5. SRU- 31/P Packets 1 and 2
6. Life preserver
7. Sheathed knife
8. Hook blade pocket knife (MC-1)
9. Plastic ball whistle
10. Signal mirror (Mk-3, Type 1)
11. Operator's manual

*12. .38 caliber special revolver
*13. .38 caliber ball cartridges
*14. .38 caliber tracer cartridges
*15. Radio

* Issued by unit commander

U.S. ARMY SARVIP VEST

The U. S. Army's newest survival vest, the Aircrew Survival Recovery Vest, Insert and Packets (SARVIP) will replace the SRU-21/P vest that has been in use for over 20 years. The new vest will incorporate armor plating, providing .50 caliber protection in the front. The new design makes the armor insert more comfortable for the pilot and the configuration of the pockets reduces bulk. In addition, the vest will add rescue lift capability and flame protection.

U.S. ARMY SARVIP VEST

The SARVIP consists of a Raschel knit Nomex, fire-resistant fabric with twelve pockets, ten outer and two inner. The vest includes a rescue lift strap, two leg straps and a chest strap. The chest strap provides a means for attaching the LPU-10/P life preserver units. The vest has two small 'D' rings attached to the front and sides for attaching the protective mask blower and carrier.

The .50 caliber armor insert is made of laminated ceramic/fiberglass and has a foam backing. The carrier for the insert is made of Nomex and seven layers of Kevlar and has a quick-release strap at the bottom for emergency release of the plate. The pockets have vacuum packed nylon-polyethylene pouches and hold all the basic individual medical survival items.

SARVIP vest (components): armor plating, LPU-10/P's (2), 2" x 3" mirror, wire saw, first aid kit, Mk-13 Mod 0 signal/flares (2), PRC-90 radio, tourniquet, pocket knife, water bladder, signal flares and launcher, SDU-5/E strobe light and cover, dye marker, lensatic compass, operators manual, chemical mask and blower, SRU-31 A/P

WORLD WAR II LIFE VESTS

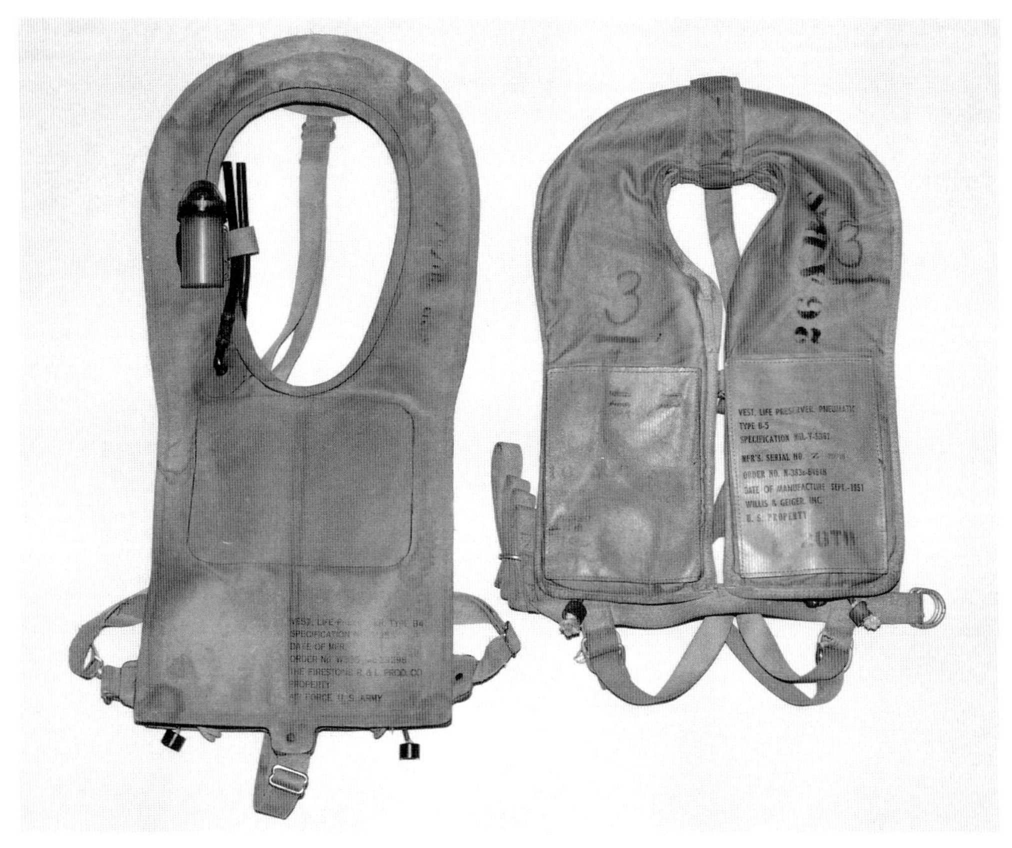

B-4 life vest (left), B-5 life vest (right)

LABEL

VEST, LIFE PRESERVER, TYPE B4
SPECIFICATION NO. 3135
DATE OF MFR.
ORDER NO. W535AC-29298
THE FIRESTONE R. & L. PROD. CO.
PROPERTY
AIR FORCE, U.S.ARMY

LABEL

VEST,LIFE PRESERVER,PNEUMATIC
TYPE B-5
SPECIFICATION MIL-V-5367

MFR'S. SERIAL NO. Z 29018
ORDER NO. N-383s-54948
DATE OF MANUFACTURE SEPT - 1951
WILLIS & GEIGER, INC.
U. S. PROPERTY

The B-4 life vest was used during World War II and into Korea and consisted of two superimposed pneumatic compartments of rubber-coated yellow fabric. The earlier B-3 life vest (not shown) was similar and was the first pneumatic LP issued to aircrews. Each compartment of the B-4 was equipped with separate mouth inflation tubes and valve assemblies, and automatic CO2 inflation systems operated by pulling respective cord. Sea dye markers are glued on either side between the folds and a shark repellent packet could substitute for one of the dye markers. Adjustable waist and crouch straps ensured proper fit.

The AN 6519 and dark blue colored U.S. Navy LP's are similar in design except for the location of the oral inflation tubes between the folds and the strap arrangement on the blue U.S. Navy LP. The Mk-1 U.S.N. life vest was a major design departure and provided better buoyancy because of its 'wrap around' body style.

WORLD WAR II LIFE VESTS

The B-5 vest was considerably different from the B-4 and AN 6519 series LP's. It provided better buoyancy, was smaller in design and more comfortable to wear. The B-5 contained two pockets and was issued with a sea dye marker packet, whistle and attachable flashlight. Shark repellent and a signal mirror could also be added. A leather patch was sewn onto either side of the cloth vest and the rubber inflation bladders were replaceable. Operation was similar to the other LP's. The B-5 vest was used during World War II, Korea and listed as limited standard into the 1960's.

AN 6519-1 life vest (left), U.S. Navy dark blue life vest (right)

LABEL

VEST-PNEUMATIC LIFE
AN6519-1
DATE OF MFG. APR 21 1945
CONT. NO. W-33-038-AC-5883
UNITED STATES RUBBER CO., N.H.
PROPERTY - U.S. GOVERNMENT

LABEL

DATE OF MANUFACTURE SEP 26 1945
CONTRACT - N151s-63707
UNITED STATES RUBBER CO.
"NEOPRENE"
PROPERTY- U.S.NAVY

WORLD WAR II LIFE VESTS

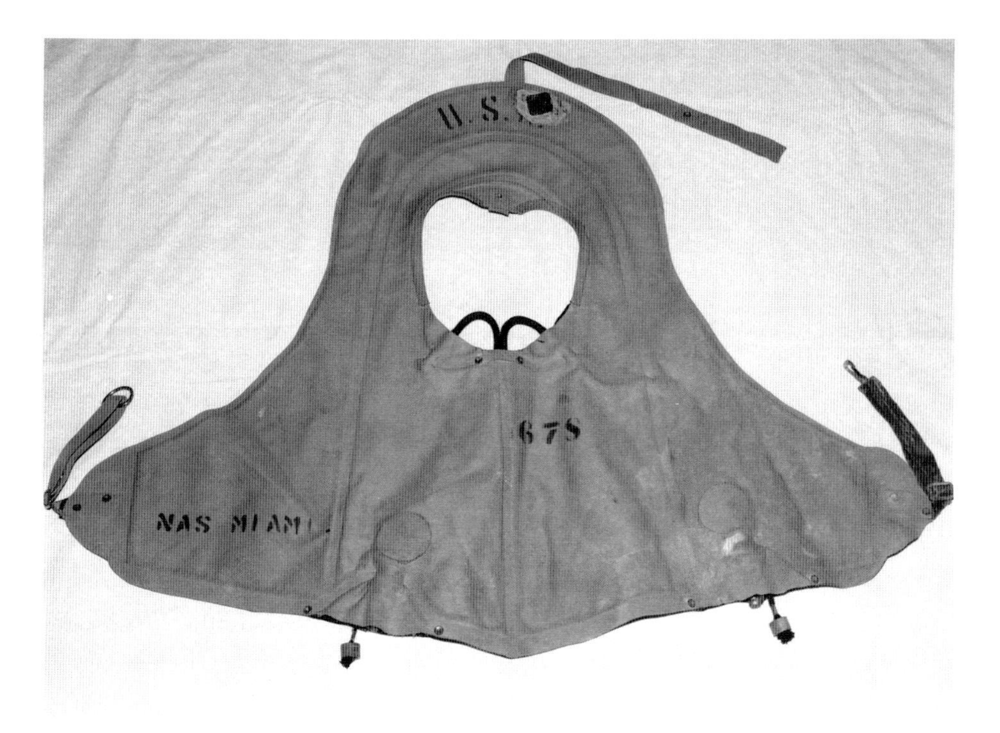

U.S.N. Mk-1 life vest

LABEL

NEW YORK RUBBER CORP.
BEACON. N. Y.
NAVY CONTRACT NO. N-288S-2883

MK-2 LIFE VEST

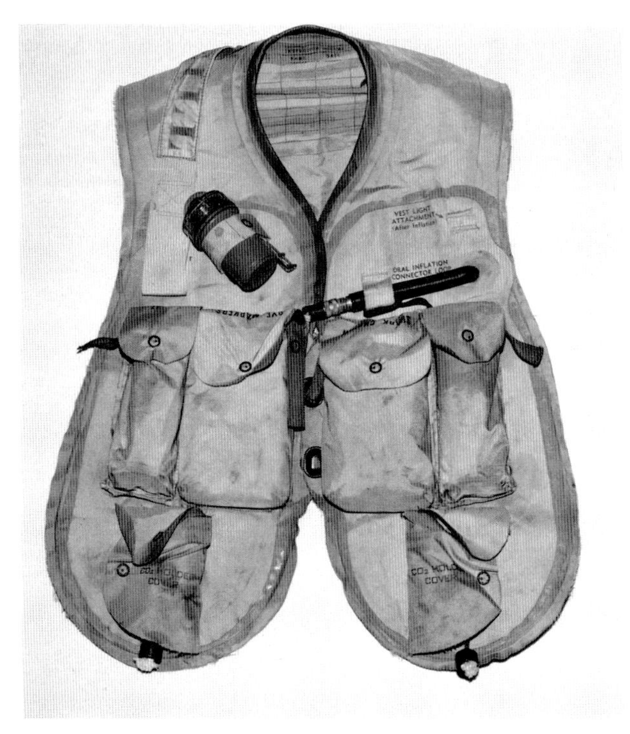

Mk-2 life vest (1957)

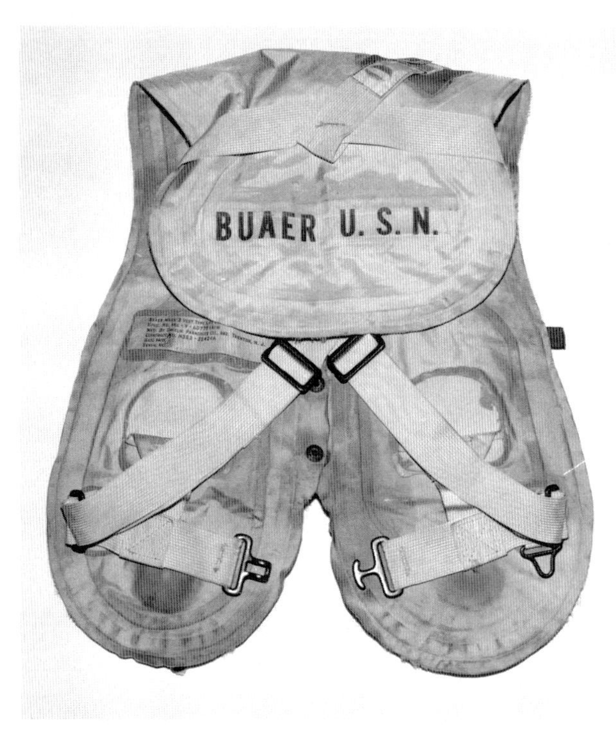

Mk-2 life vest (back)

LABEL

BUAER MARK 2 VEST TYPE LIFE PRESERVER
SPEC. NO. MIL-V-60778(AER)
MFD. BY SWITLIK PARACHUTE CO., INC. TRENTON, N.J.
CONTRACT NO. N383-25424A
DATE MFD. JAN 1957
SERIAL NO. 15320

The Mk-2 assembly is a vest type preserver constructed of chloroprene-coated nylon cloth and consists of a three compartment floatation assembly, a harness and waist strap assembly and two inflation assemblies. Four outer pockets hold five inner pouches containing the survival equipment and attached to the ouside of the vest is a distress light. Early Mk-2 vests did not have pockets for the dye markers and shark repellent, these items being tied to small loops on the vest. A 'D' ring is incorporated into the vest for attachment to the lanyard of the raft accessory kit. A whistle and compass are sometimes added to the vest and are attached by a lanyard to the oral inflation tube retaining strap.

The Mk-2 vest is manually inflated by pulling both assembly lanyards, attached to CO2 cylinders, down simultaneously. The Mk-2 is not designed for inflation through the oral inflation valves but in an emergency situation the center compartment may be inflated through the oral inflation valve to provide additional buoyancy.

The Mk-2 vest has been in use from the late 40's thru Vietnam and was still listed in Air Force T.O.'s in the late 70's. Earlier models were yellow in color and marked BUAER U.S.N. on the back. Later vests were orange and marked BUWEPS U.S.N. Many vests were also dyed black during Vietnam.

MK-2 LIFE VEST

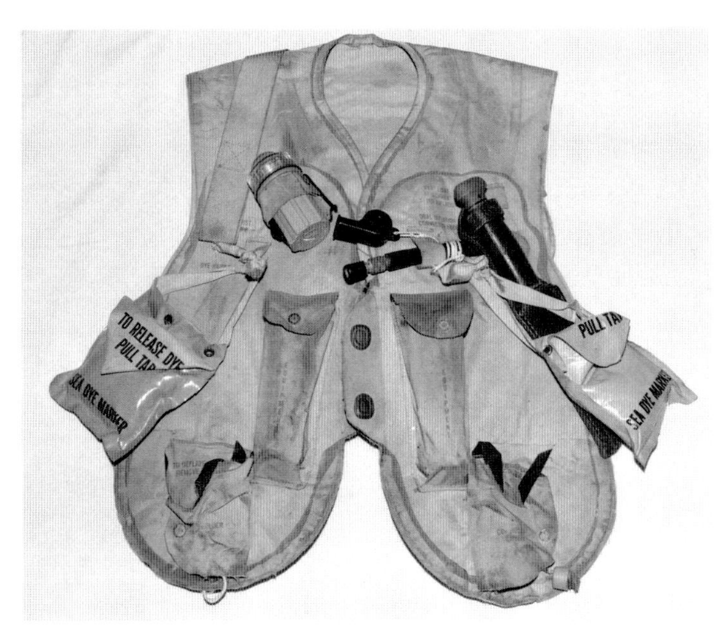

Mk-2 life vest (1952)

LABEL

BUAER MARK 2 LIFEVEST
BUAER SPEC. 23P17(AER)
MFD. BY WILLIS & GEIGER INC.
CONTR. NO. N383S-38767
DATE MFD. MAR 1952

LABEL

BUWEPS MARK 2 INFLATABLE VEST LIFE PRESERVER
SPEC. NO. MIL-L-6077G(WEP)
MFD. BY SWITLIK PARACHUTE CO., INC. TRENTON, N.J.
CONTRACT NO. N383-96599A
DATE MFD. FEB 1969
SERIAL NO. 126549

Mk-2 life vest (1969)

MK-2 LIFE VEST

Components for the Mk-2:

- -Distress signal light (SDU-5/E strobe on later models)
- -Mk-13 Mod 0 smoke/flares (2)
- -Dye markers (2)
- -Shark repellent
- -Whistle
- -Hunting knife
- -L-1 or MC-1 compass

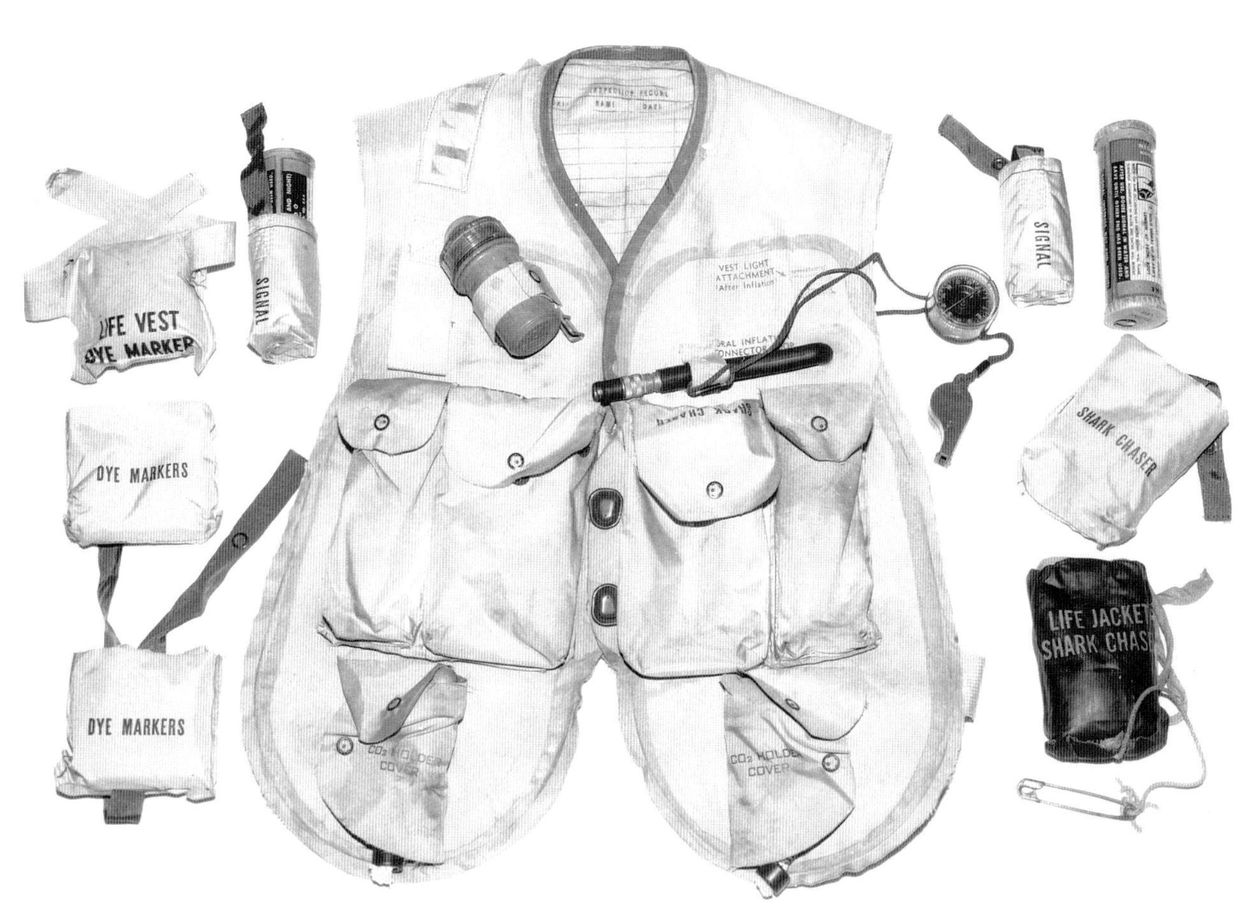

Mk-2 life vest components (left to right): dye markers (2), Mk-13 Mod 0 smoke/flares (2), shark chaser, whistle, MC-1 compass, signal light pinned on vest

MK-3C LP

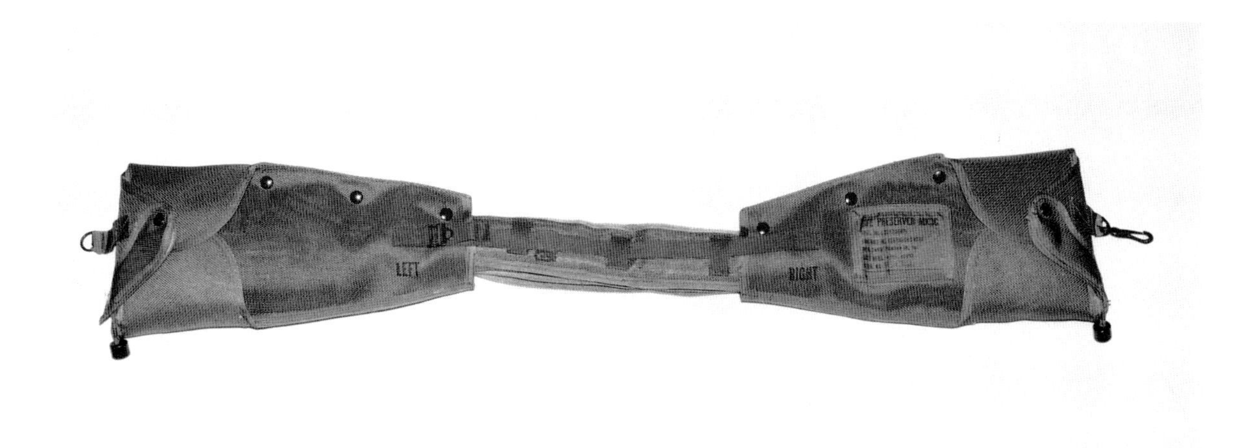

Mk-3C life preserver (1970)

<u>LABEL</u>

LIFE PRESERVER MK-3C
SPEC. MIL-L-22276A(WP)
CONTRACT NO. DSA 700-69-C-H223
MFR. SWITLIK PARACHUTE CO., INC.
DATE MFRD. APR 1970
SERIAL NO. 43132

The U.S. Navy Mk-3C life preserver was designed to be worn with the integrated torso harness and could be used with the SV-1 and SV-2 survival vests. This LP was first employed in the late 1950's and was the predecessor to the LPA-1 'horse collar' life preserver. It is worn around the waist of the wearer and had two compartments, and upper and lower, in the shape of a 'C', with the opening in front of the pilot. The two ends were connected by means of a snap fastener. The LP was inflated by pulling an inflation lanyard located below and in front of each bladder compartment and attached to a CO2 cylinder.

In addition, the Mk-3C could be manually inflated through two oral inflation valves. Two different bladder colors have been seen, yellow and orange. Pockets were provided for Mk-13 smoke/flares, dye markers or shark repellent. This life preserver was still being used into the 1970's.

LPA-1 AND LPU-23/P LP'S

LPU-23/P 'Horse Collar' life preserver showing attached dye marker and flare pouches.

LABEL

LIFE PRESERVER
LPU-23B/P
SERIAL NO.
ACTIVITY

The U. S. Navy LPA-1 was the first in a series of 'Horse Collar' design life preservers, progressing through to the LPU-23/P in use today. The term 'Horse Collar' applied because of the design of the LP. The lower section was worn around the waist and an upper section worn around the neck. Both sections were connected by an inflation channel that ran up the back. These LP's integrated into the SV-2 survival vests through two waist straps and two 'D' rings on the SV-2.

The dark blue bladders were inflated manually on the LPA-1 and manually or automatically on the LPU-23/P. Manual inflation on the LPA-1 was accomplished by pulling two inflation lanyards connected to C02 cylinders. The LPU-23/P was manually inflated in a similar fashion except the lanyards incorporated a series of six small round balls on each lanyard to aid in tactile feel. The LPU-23/P can also be automatically inflated through the use of two 'automatic inflation' devices, designated FLU-8/P. These devices will automatically inflate the LP upon contact with salt water. Three 'D' rings are positioned on each lower side of the LP's for attachement of a Mk-13 smoke/flare pouch and a sea dye marker pouch. Early pouches had the words 'flares' and 'dye marker' stenciled in yellow and later ones were in stenciled in black.

LPA-1 AND LPU-23/P LP'S

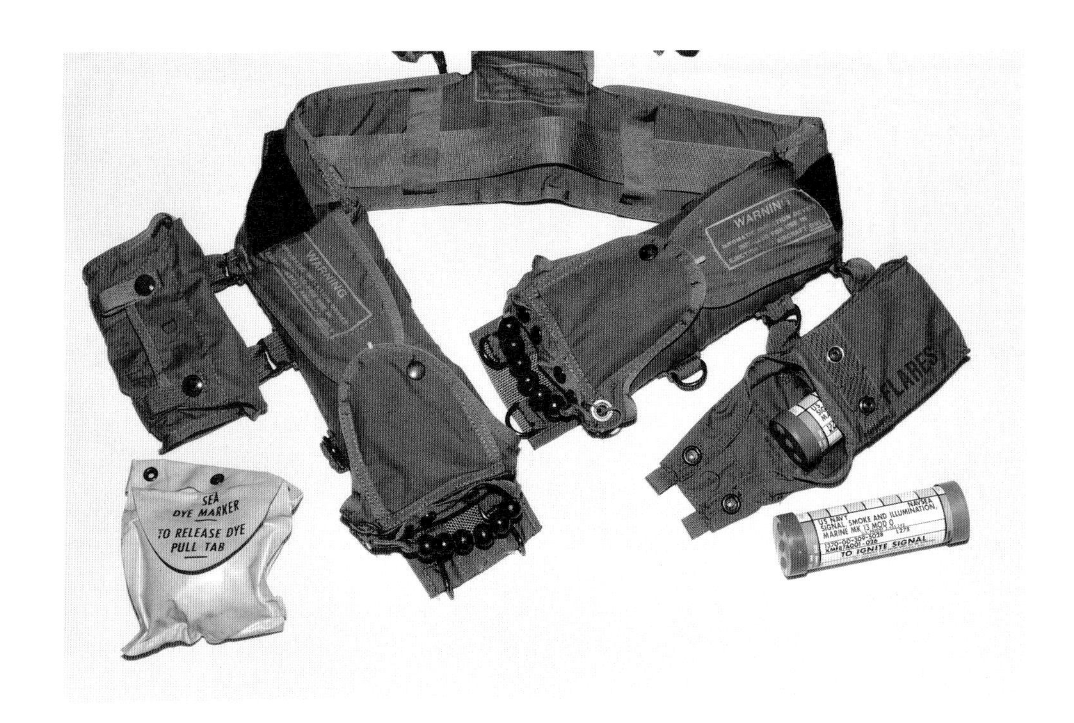

Left: LPU-23/P life preserver showing dye marker and flare pouch details

LABEL

CASING, LIFE PRESERVER TYPE LPA-1
SPEC. MIL-L-81561(AS)
CONTRACT NO. DSA 700-70-C-9561
SWITLIK PARACHUTE CO., INC.

Above: LPA-1 life preserver

MA-2, LPU-2/P, LPU-3/P, LPU-9/P AND LPU-10/P LP'S

LPU-2/P life preserver

LABEL

LIFE PRESERVER, UNDERARM, PNEUMATIC LPU-2/P
MIL-L-26558B(USAF)
RUBBER FABRICATORS, INC.
CONT. NO. N 383(MIS)93637A
DATE MFG. AUG. 1966

The MA-2, LPU-2/P, LPU-3/P and the LPU-10/P life preservers consist of two orange colored neoprene-coated floatation cells, each packed into a flat envelope type container. The MA-2, 2/P, and 10/P containers are attached to an adjustable harness which secures the preserver to the body. The 3/P containers are similar to the ones used on the 2/P and 10/P, but in lieu of a harness, they are attached directly to the parachute harness by means of slide fasteners. Each cell is inflated mechanically by a C02 cylinder. Pulling downward and outward on the lanyards that extend from the lower front corner of each container inflates the preserver. If a failure occurs, the life preserver may be orally inflated through the oral inflation tube at each cell.

The MA-2 was the earliest style of this type of LP and was the replacement for the B-5 life vest. The MA-2 container measured 8" x 5" x 1 1/4" and had provisions for attaching a zippered accessory packet to each cell. These packets could contain sea dye markers, signal mirrors, etc. The LPU-2/P replaced the MA-2 because it was thinner in the packed configuration and was designed to vent air in the cells at altitude or during decompression.

The 2/P and the 10/P are identical except for the type of inflator used (MA-1 inflator on the 2/P and the FLU-1/P on the 10/P and 3/P). These life preservers have been used primarily by the Airforce and Army.

The current issue U. S. Air Force life preserver is the LPU-9/P. This LP is worn around the neck and chest. It attaches to a parachute torso harness with a slide fastener behind the neck and a clip fastener attached to a 'D' ring on each side of the harness. The dark blue floatation bladder is encased in green Nomex cloth secured by Velcro. The 9/P can be activated manually or automatically through an 'automatic inflation' device designated as the FLU-9/P that, upon contact with salt water, inflates the LP. This is a particularly important feature for an unconscious pilot landing in the water. A similar automatic inflation device has been and is still being used by the Navy in their LPU-23/P life preservers.

MA-2, LPU-2/P, LPU-3/P, LPU-9/P AND LPU-10/P LP'S

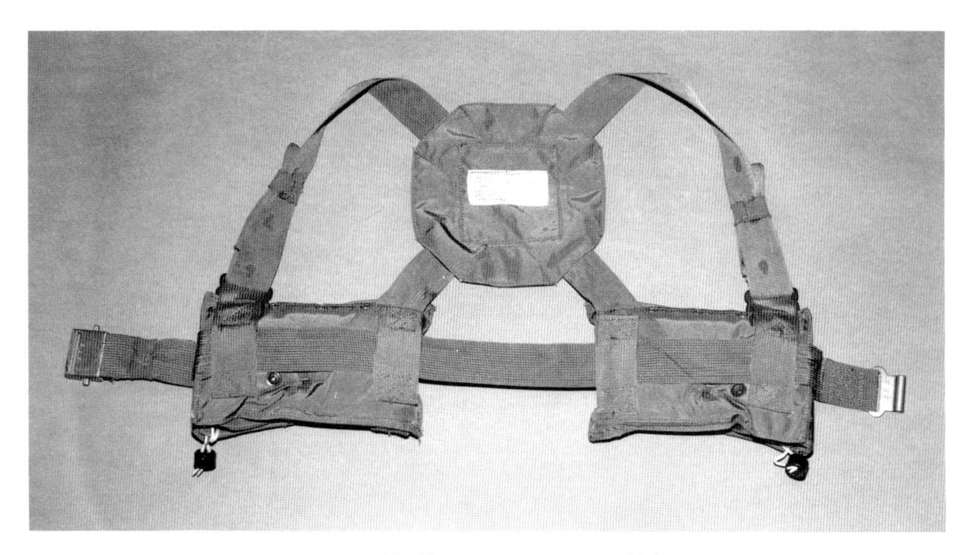

LPU-10/P life preserver (reverse side)

LABEL

LIFE PRESERVER, UNDERARM
PNEUMATIC, LPU-10/P
SPECIFICATION MIL-L-38484(USAF)
RUBBER CRAFTERS OF W. VA., INC.
CONTAINER PART NO. 58K3589
CONTRACT NO. F41608-82-C-187
DATE OF MANUFACTURE SEPT 1982

MA-2 life preserver and accessory pouches
(1/2 of life preserver missing)

MA-2, LPU-2/P, LPU-3/P, LPU-9/P AND LPU-10/P LP'S

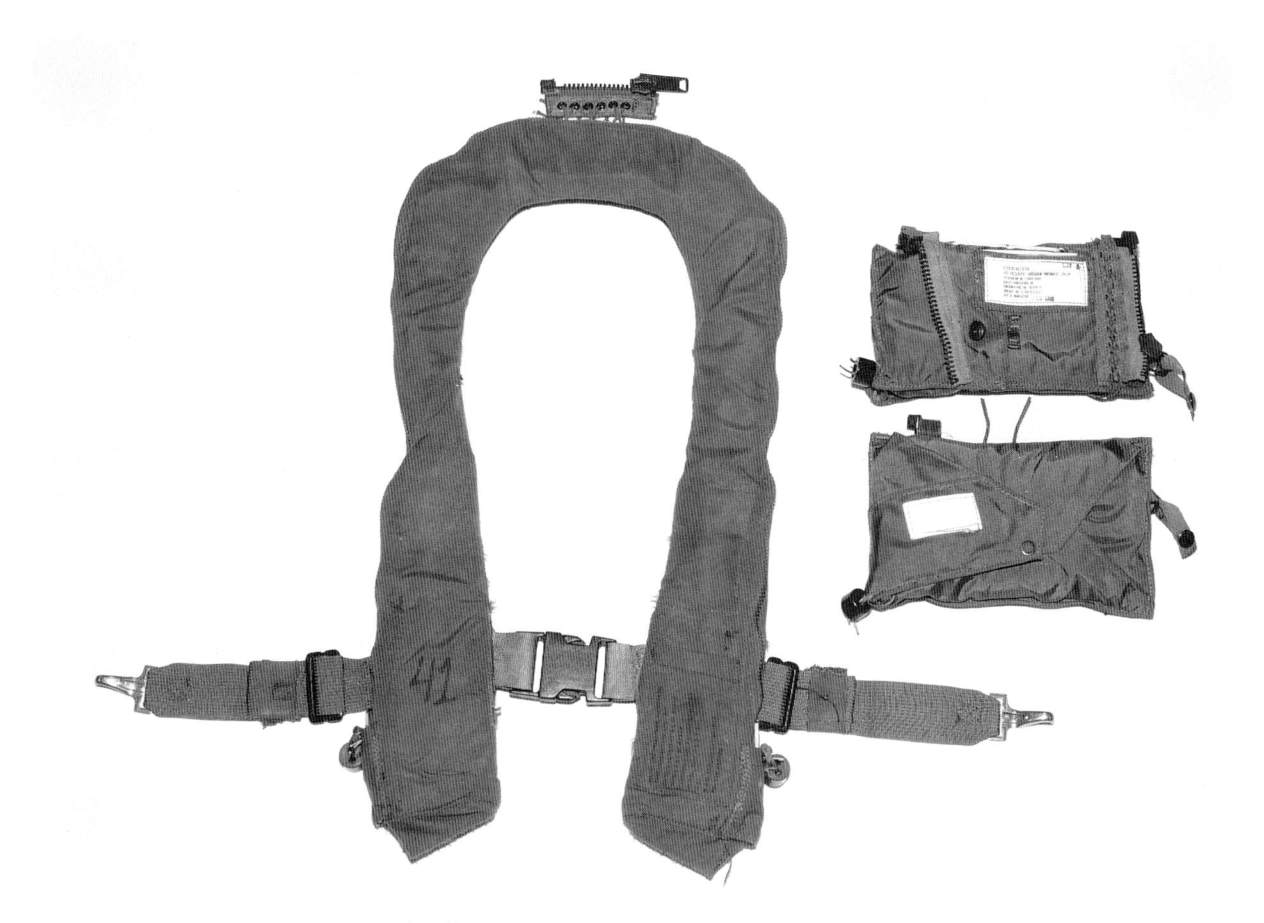

LRU-9/P life preserver (left), LRU-3/P life preserver (right)

<u>**LABEL**</u>

NAME:LIFE PRESERVER,AUTOMATIC
TYPE DESCRIPTION:LPU-9/P
DESIGN ACTIVITY F S C M:99440
MFG'S F S C M:62323
CONTRACT NO:F33657-87-C-0197
SERIAL NO:
U.S.:
NSN:4220-01-051-5916LS
P/N:025-850100-1

<u>**LABEL**</u>

LIFE PRESERVER, UNDERARM
PNEUMATIC, LPU 3/P
SPECIFICATION #MIL-L-38484(USAF)
F.T.S. CORP.
CONTAINER PART NO. 58J3590-1L
CONTRACT NO. F41608-85-C-1435
DATE OF MANUFACTURE DEC 1985

CHAPTER 2

SURVIVAL SEAT AND BACK PAD KITS

Survival seat and back pad kits have been in consistant use since the advent of the one man liferaft container kits issued during World War II. The seat kits consisted of a one- man liferaft and a few survival items contained in a canvas bag attached to a parachute harness. The bag also doubled as a seat cushion for the pilot. Many types of seat and back type kits have been developed since World War II and can generally be categorized as to the major environment of use, i.e. overwater (containing a liferaft), hot climate and cold climate kits.

The AN-R-2 and C-2 series liferaft kits were among the first overwater kits issued during World War II and could be supplemented by other types of survival kits worn or carried by the pilot or crewmember. The M-592, B-1, B-2 and B-4 back/seat kits and the C-1 survival vest were used with the raft kits and contained items usually not included in the raft container.

In the 1950's even more development and specialized containers were used with the ejection seats in the fighter and bomber aircraft. The ejection seat containers were made of either combinations of nylon and fiberglass or magnesium or all fiberglass. The term 'hard kit' described the all fiberglass containers vs. 'soft kit' for containers made of nylon or canvas. The survival kit components were based on the mission requirements and whether or not a survival vest was worn.

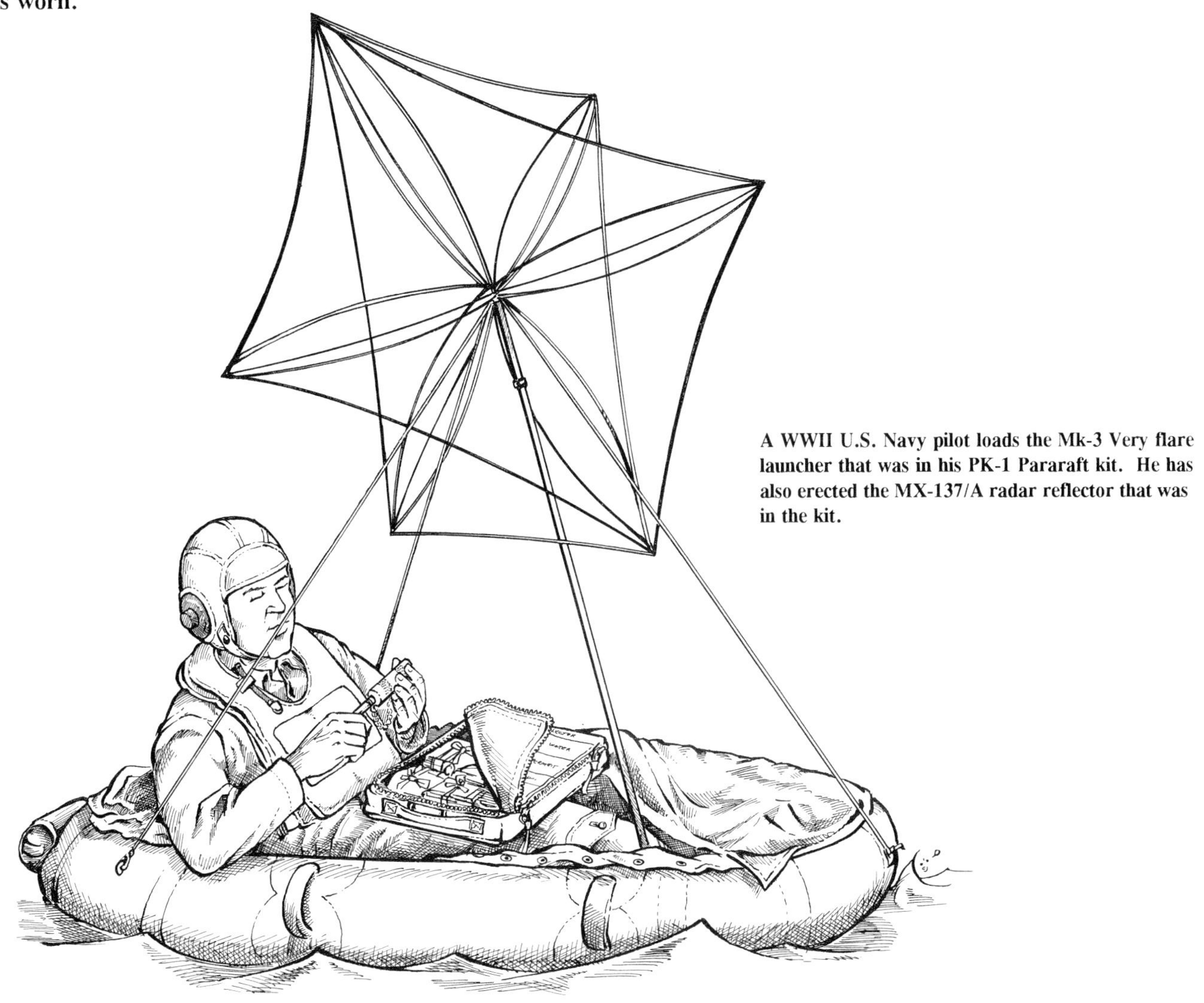

A WWII U.S. Navy pilot loads the Mk-3 Very flare launcher that was in his PK-1 Pararaft kit. He has also erected the MX-137/A radar reflector that was in the kit.

AN-R-2 RAFT KIT

AN-R-2A raft kit

LABEL

LIFE RAFT
AN-R-2A
RIP OPEN COVER
PULL OUT LOCK PIN
TURN VALVE TO LEFT

TESTED MO.A.S.O.
9-2-44

LABEL

ORDER NO. 43-23469-AF
U.S. RUBBER CO.
MFG. 9/4/43
INSPECTED 9/8/43

The AN-R-2 was one of the first one man parachute liferaft containers to be issued during World War II and can be used with either the seat type or the quick attachable chest type parachute. The yellow-orange colored case measures 15" x 14" x 4" and weighs about 13 1/2 pounds. A closure flap is secured by a series of snaps on four sides.

When used in conjunction with the seat type parachute, it is stored between the parachute and the body in place of the parachute seat cushion. An opening located midway between the center and forward edge permits the leg straps of the parachute harness to pass through the container. Two non-ejector snaps on the container attached it to parachute 'D' rings and a lanyard connected the kit to the aircrew member's LP.

When used with the quick attachable chest type parachute, it is stored in a case located between the parachute and body. There are three labeled compartments within the container for the survival items and each compartment has a lift-the-dot snap fastener to secure its flap.

AN-R-2 RAFT KIT

The AN-R-2a contains a sail whereas the AN-R-2b does not.

The raft is inflated by first unsnapping the closure flap of the kit and removing the raft and CO2 cylinder. A safety pin was removed from the CO2 cylinder valve handle and the valve handle turned to allow inflation of the raft.

Components for the AN-R-2 kit are:

- -AN-R-2 series raft (yellow-orange in color) and sea anchor
- -Bailing cup
- -First aid kit (lifeboat/liferaft type)
- -Raft repair kit
- -Paddles (2) (blue canvas covering a wire form with arm straps)
- -Leak plugs (4)
- -Emergency can of drinking water
- -Dye marker (in a metal can)
- -Paulin
- -Sail

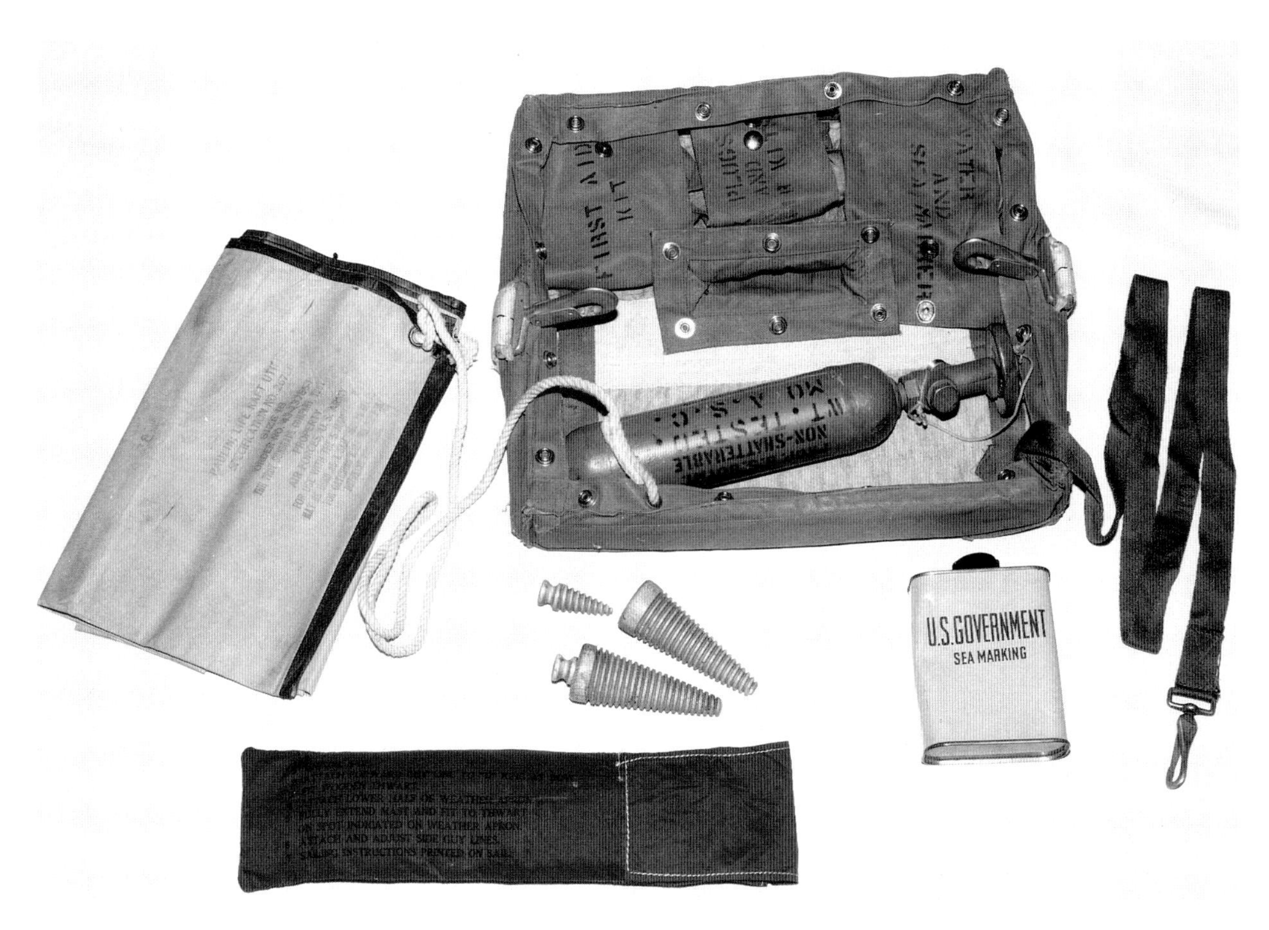

AN-R-2 components (partial): paulin, 'bullet' hole plugs, sail, sea marker, CO2 cylinder (missing water can, first aid, raft)

C-2 RAFT KIT

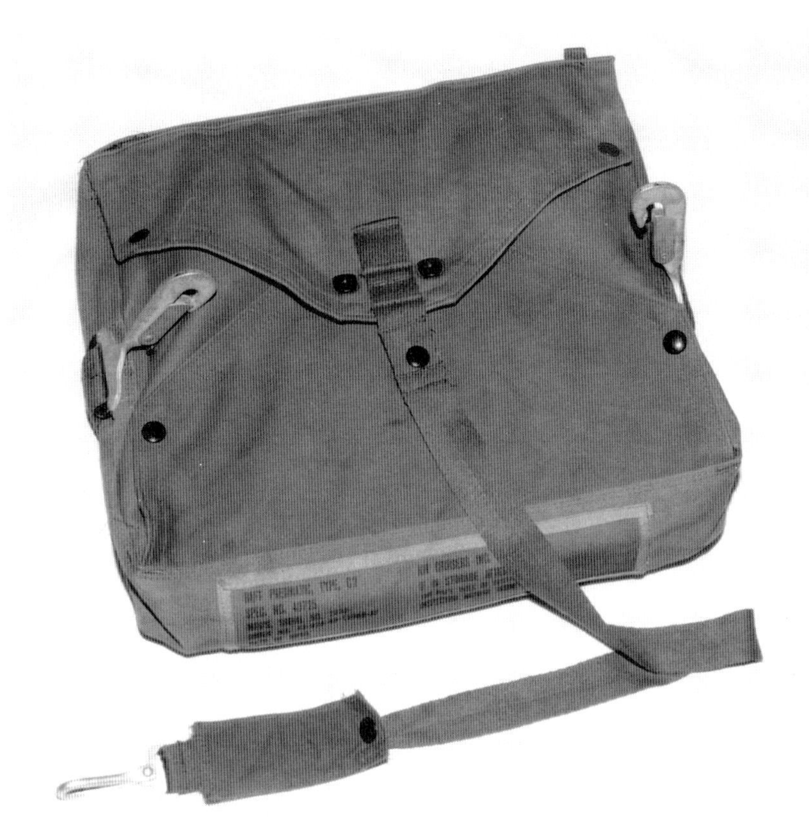

C-2 raft kit

LABEL

RAFT PNEUMATIC TYPE, C2
SPEC. NO. 40725
MFG'S SERIAL NO.
CONTRACT NO. W33-038-AC-11049
DATE OF MFGR. AUG 1945

AIR CRUISERS, INC.

The C-2 raft kit was a modification of the earlier AN-R-2a raft with principal changes being the addition of a spray shield, sail and mast, changes in accessories and packing in an improved container. The 15" x 13" x 3 1/2" outer container uses a removable inner container that also functioned as part of the raft outer securing flap. The inner compartment has three sections, each secured by a snap flap, and retaining straps for paddles. Two non-ejector snap hooks secured the C-2 to the parachute harness. A 30" long lanyard and spring clip are attached to the inner container to secure it to the 'D' ring on the life preserver. A carrying handle was located on the bottom of the container.

The raft is inflated by unsnapping the two piece cover flaps, removing the raft and pulling the inflation lanyard on the CO2 cylinder.

C-2 RAFT KIT

Components for the C-2:

- -C-2 raft and sea anchor
- -Bailing cup
- -Metal repair plugs
- -Sponge
- -Vinylite containers for storage of accessories or water
- -Rust preventative container
- -Collapsible aluminum mast
- -Sail with rigging and container
- -Paddles, wooden
- -MX-137/A radar reflector or AN-CPT-2 radar beacon
- -Signal mirror, ESM-1
- -Desalting kit
- -Spray shield
- -Distress signal, 2-star, red, hand held

The sailing equipment would be removed if the corner reflector or the radar beacon is used. The AN-CPT-2 radar beacon is a battery operated beacon measuring 15" x 4 1/2" x 2" and weighing less than three pounds. It can operate up to eighteen hours.

The corner reflector is a device to increase the distance at which life rafts could be picked up by radar. It stores in a 14 1/2" x 2 1/4" x 1 1/2" container and upon removing is assembled, erected and secured to the raft. The MX-137/A is used with one man rafts and the MX-138/A is used with the multi-man rafts.

Many C-2 raft kit containers were hand stamped with an 'A' following the C-2 designation. These were used concurrently with the later C-2A raft kits until supplies of the newer kits became available.

C-2 inner container

C-2A RAFT KIT

C-2A raft kit

LABEL

RAFT PNEUMATIC TYPE C-2A
SPEC. NO. MIL-R-5863A Amdt.1
MFR'S. SERIAL NO.
CONTRACT NO. N383s-58829
DATE OF MFR.

MFRD. BY NEW YORK RUBBER CORP.

The C-2A raft kit evolved from the C-2 kit and was used in late World War II and into the 50's. It is very similar to the ML-4 raft kit that was developed later. The o.d. C-2A container measures 15" x 13" x 4" and contained the C-2A raft and a separate inner tan colored container (accessory kit) for the survival equipment. The raft and inner container are held into the main compartment by a flap that is secured by snaps on two sides and a zipper closure across the front.

The flap is padded for use as a seat cushion and the outer container has a carrying handle attached at the front. The 14 1/2" x 4" x 3" inner container has a top zipper closure and two yellow cloth securing lines sewn to it. One line was 29" long and has a snap hook on the end to secure it to the life preserver. The second line is 65" long and is used to attach the inner container to the C02 cylinder of the raft. The MX-137/A radar reflector is tied to the top of the inner container. Two non-ejector type snap fasteners were used to attach the container to the parachute 'D' rings.

The raft is inflated by pulling a lanyard attached to the zipper keeper on the front of the bag and the raft C02 cylinder. Once the zipper keeper is removed the inflation of the raft will open the remaining portion of the container flap.

C-2A RAFT KIT

Components for the C-2A:

- -C-2A raft (yellow and pale blue in color) and sea anchor
- -Blue wooden paddles (2)
- -Mk-13 Mod 0 smoke/flares (2 or 4) (Mk-1 Mod 0, Mk-1 Mod 1 smoke/flares or M-75 'red star' distress signals were used earlier.)
- -Sponge
- -Raft repair kit
- -Mk-2 desalting kit
- -3" X 5" signal mirror (B-1, ESM-1 or ESM-2)
- -Vinylite containers for water or accessories (2)
- -Container of rust preventative
- -Bailing cup
- -MX-137/A radar reflector
- -Morse code card (Not installed when M.C. is stenciled on the raft floatation tube.)

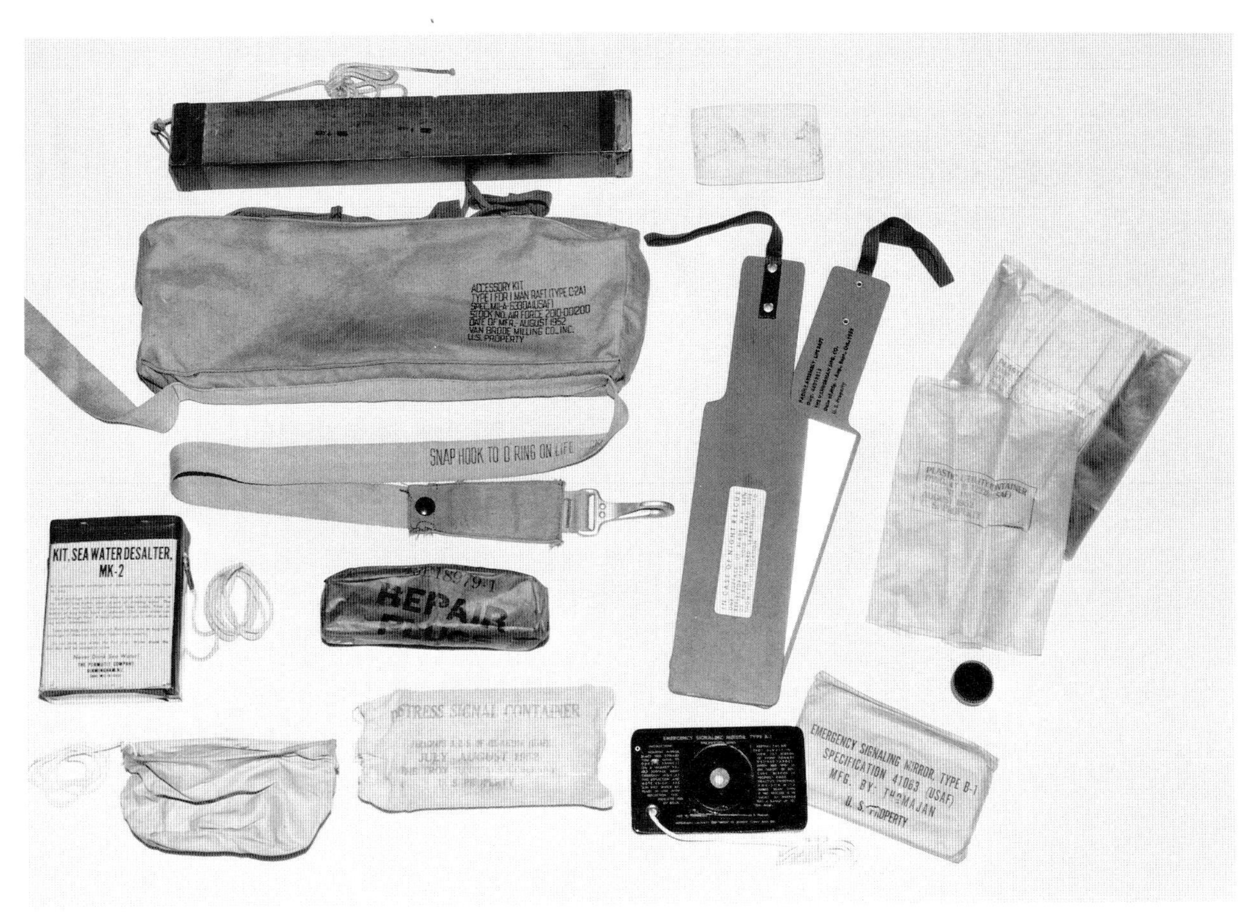

C-2A kit components (left to right): radar reflector, accessory kit container, desalter kit, bailing bucket, Mk-13 Mod 0 smoke/flares in pouch, raft repair kit, B-1 3" x 5" mirror, paddles, sponge, vinylite containers, rust preventative

C-2A RAFT KIT

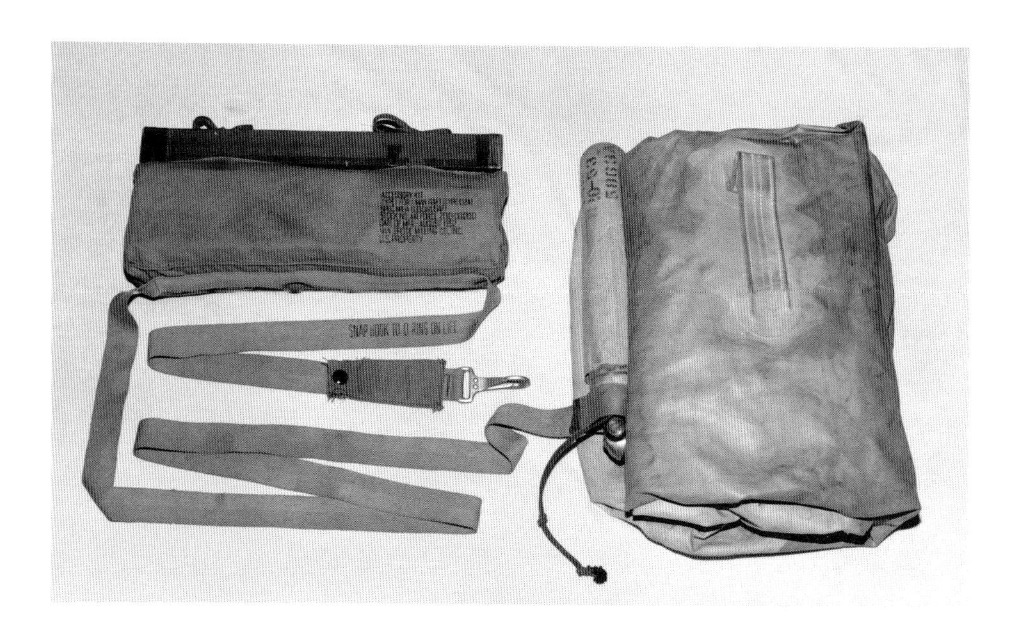

C-2A kit raft and accessory kit container

ACCESSORY KIT LABEL

ACCESSORY KIT
TYPE 1 FOR 1 MAN RAFT(TYPE C-2A)
SPEC. MIL-A-6330A(USAF)
STOCK NO. AIRFORCE 2010-001200
DATE OF MFR. - AUGUST 1952
VAN BRODE MILLING CO., INC.
U.S. PROPERTY

RAFT LABEL

RAFT, PNEUMATIC, TYPE C-2A
SPEC. NO. MIL-R-5863A
NEW YORK RUBBER CORP.
CONTRACT NO. N383-S-75341
MFGR'S SERIAL NO.
DATE OF MANUFACTURE 1952

PK-1 PARARAFT KIT

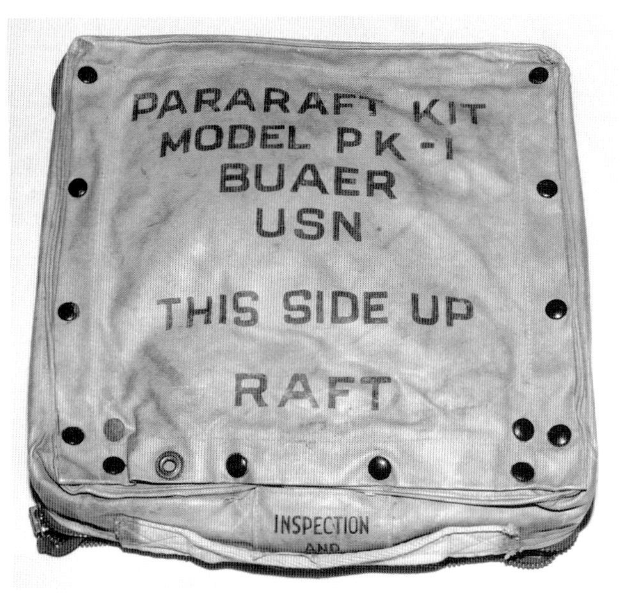

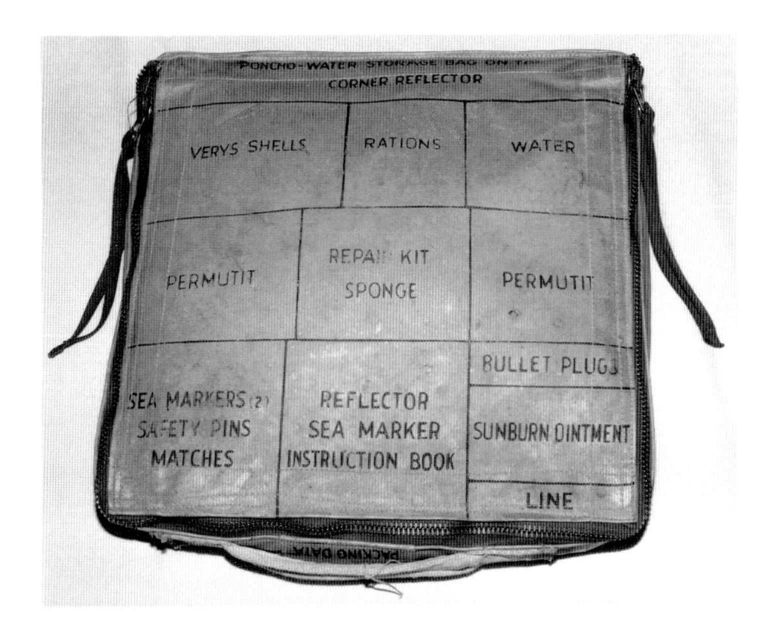

PK-1 Pararaft Kit (raft side)

PK-1 Pararaft Kit (equipment side)

During World War II, the Navy used the PK-1 Pararaft Kit in a similar fashion to the AN series of one-man life raft containers. The 'Model A' Pararaft Kit was a limited production design that preceeded the PK-1 and was employed until the PK-1 became available. It was similar in design to the PK-1 and used components from the M-592 Back Pad Kit and AN-R-2 series. The yellow container measures 15" x 14" x 3" and has a top compartment for the one-man raft and a bottom compartment for the survival equipment. The flap for the raft compartment is secured with a series of snaps around 3 sides. The equipment compartment flap uses a slide fastener around three sides. A carrying handle is attached to the front of the container. The equipment flap labeled each location of all the survival components contained inside and each compartment had cotton ties which secured each item in place. The raft is inflated by first opening the raft compartment flap, removing the folded raft, and opening the hand valve on the CO2 raft bottle until the raft was fully inflated.

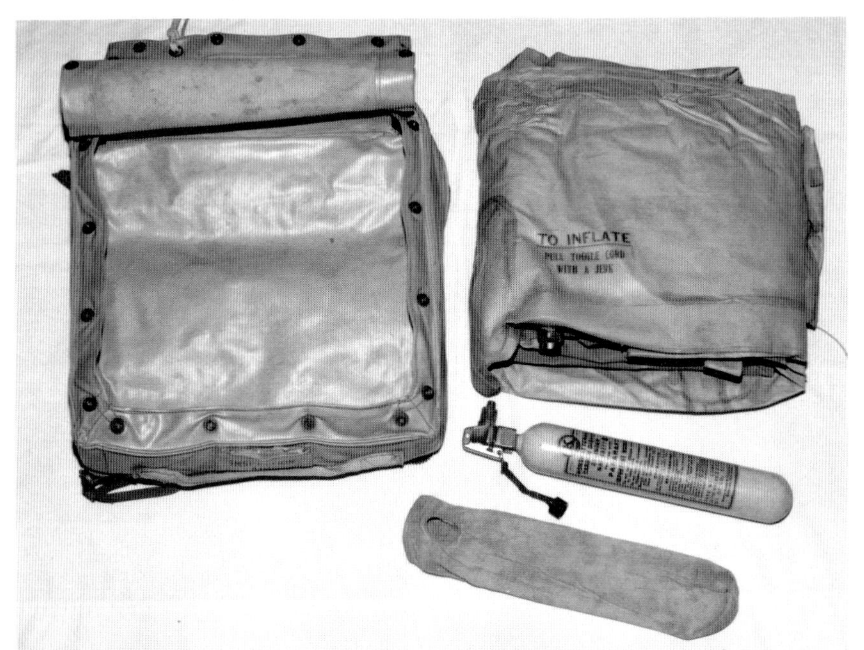

RAFT LABEL

PARARAFT

NAVAER SPEC. M-641
AIR CRUISERS, INC.
CONTRACT NO. NOA(S)5293
DATE OF MANUFACTURE JUL 1945

Left: Raft side open to show folded raft, CO2 cylinder, CO2 cylinder protective case

PK-1 PARARAFT KIT

Components for the PK-1 are:

-PK-1 raft (yellow/light blue in color) and CO_2 cylinder with sea anchor
-Poncho
-Desalting kits (2 cans)
-Can of water
-Raft repair kit
-Sponge
-Sea dye marker
-M-590 or ESM-1 signal mirror
-Instruction book
-Sunburn ointment
-Wooden bullet plugs (2 small, 2 large)
-U.S.N. tablet rations (3 cans)
-M-3 or M-4 Very shell projector and 6 Very shells
-Radar reflector
-Water storage bladder
-Waterproof match case and matches
-Safety pins (5)
-Paddles (2) (either wire covered blue canvas or wooden)
-Combination hat/mosquito headnet
-Sail
-25 foot cotton line

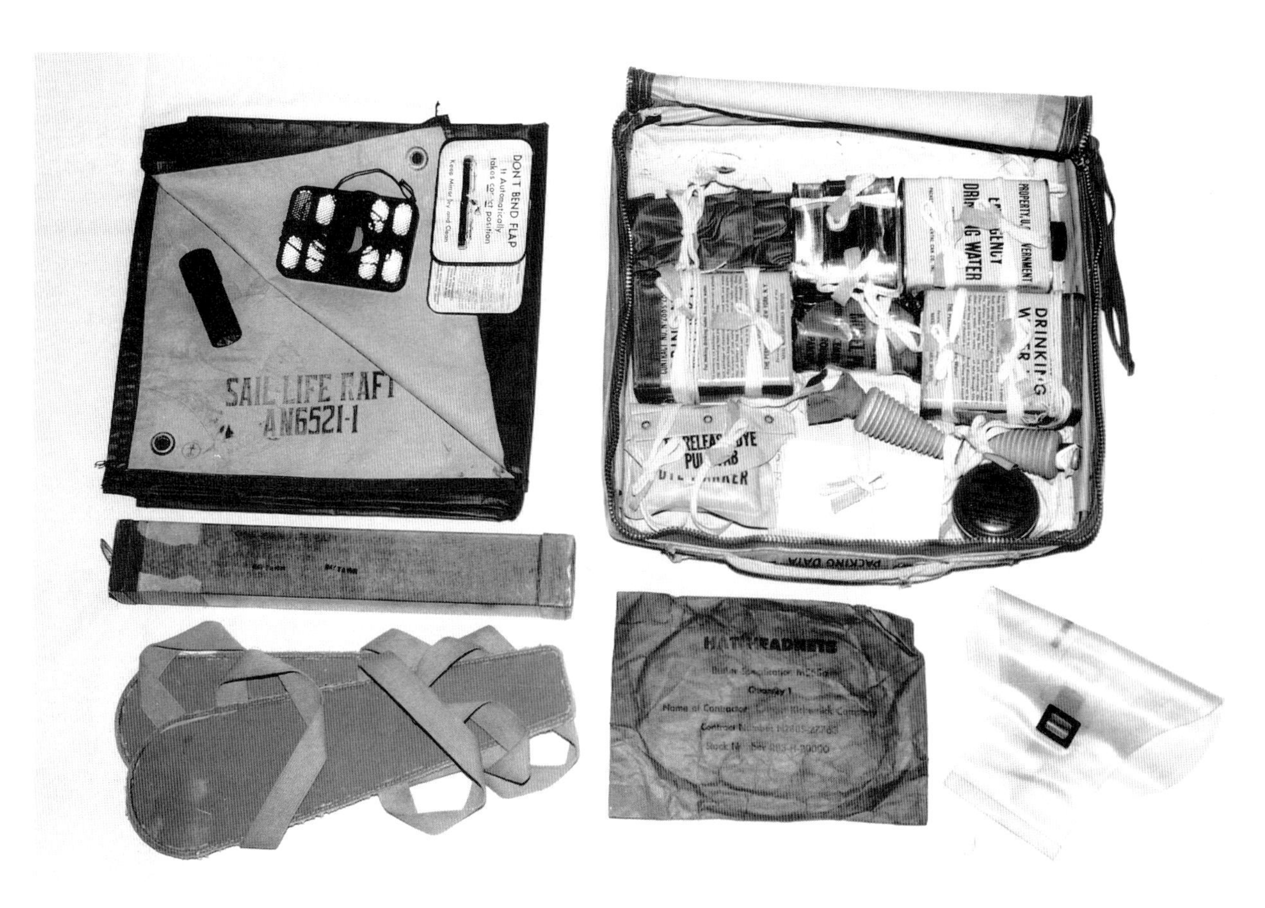

PK-1 kit components (left to right): paddles, radar reflector, sail, match case, M-590 mirror, hat/mosquito headnet, 3 pint water bladder, sunburn cream, bullet hole plugs, can drinking water, sea dye marker, desalter kits (2), raft repair kit, U.S.N. tablet rations (3), Very shells and projector in bag, sponge

M-592 BACK PAD KIT

U.S.N. M-592 Back Pad Kit, earlier variation

The M-592 Back Pad Kit was used by the U. S. Navy in World War II and followed previous designs resembling the Army Air Force B-1 and B-2 back pad kits. The M-592 was not attached to the parachute harness but was designed to be worn as a back pack under the life vest and parachute harness. Adjustable shoulder straps were provided to secure the unit to an individual. The kit was also used with the AN-R-2 and PK-1 Pararaft Kit. The M-592 was rectangular in shape, measuring approximately 19" x 11" x 4 1/2", grey in color and had three minor variations as far as nomenclature and container material. Early kits were made of a waterproof canvas and later ones had a smooth surface fabric. It was opened/closed by a slide fastener on three sides and the components were stored in specifically marked pockets with snap fasteners or by cotton snap ties. The majority of components were wrapped and sealed in foil lined brown paper.

Components for the M-592:

<u>Inside Top Cover:</u>

- -Mosquito headnet M-565
- -Cotton gloves
- -Blue/yellow poncho that could attach to the life raft and be used as a spray shield

<u>Center Compartment:</u>

- -Ten inch bladed non-folding machete with blade guard and leather wrist thong (blue steel blade, wooden grip)
- -Instruction booklet and pencil
- -ESM-1 signal mirror
- -Magnifying/burning lens

M-592 BACK PAD KIT

Components for the M-592 (continued)

Bottom Compartment:

-First aid kit, 6 unit
  4" compress bandages (2 boxes)
  Sulfanilamide (1 box)
  Sulfadiazone (1 box)
  Morphine syrettes (6) and iodine in 1 box
  Burn ointment (2 tubes) and seasick pills in 1 box

-Fishing kit AN-L-2
-Whistle M-592
-Match case/compass with matches (string lanyard attached)
-Emergency drinking water (2 cans)
-Sharpening stone (4" x 1" x 1/4")
-U.S.N. Tablet Rations (3 tins)
-Sunburn ointment
-Corrosion preventative (tube of Kant-Rust or can of oil)
-M-3 Very projector and 10 ga. Very shells, red (6)
  (Very shells were individually sealed in plastic tubes.)
-25 foot cotton line
-Container of safety pins, salt tablets and adhesive tape
-Jackknife M-575

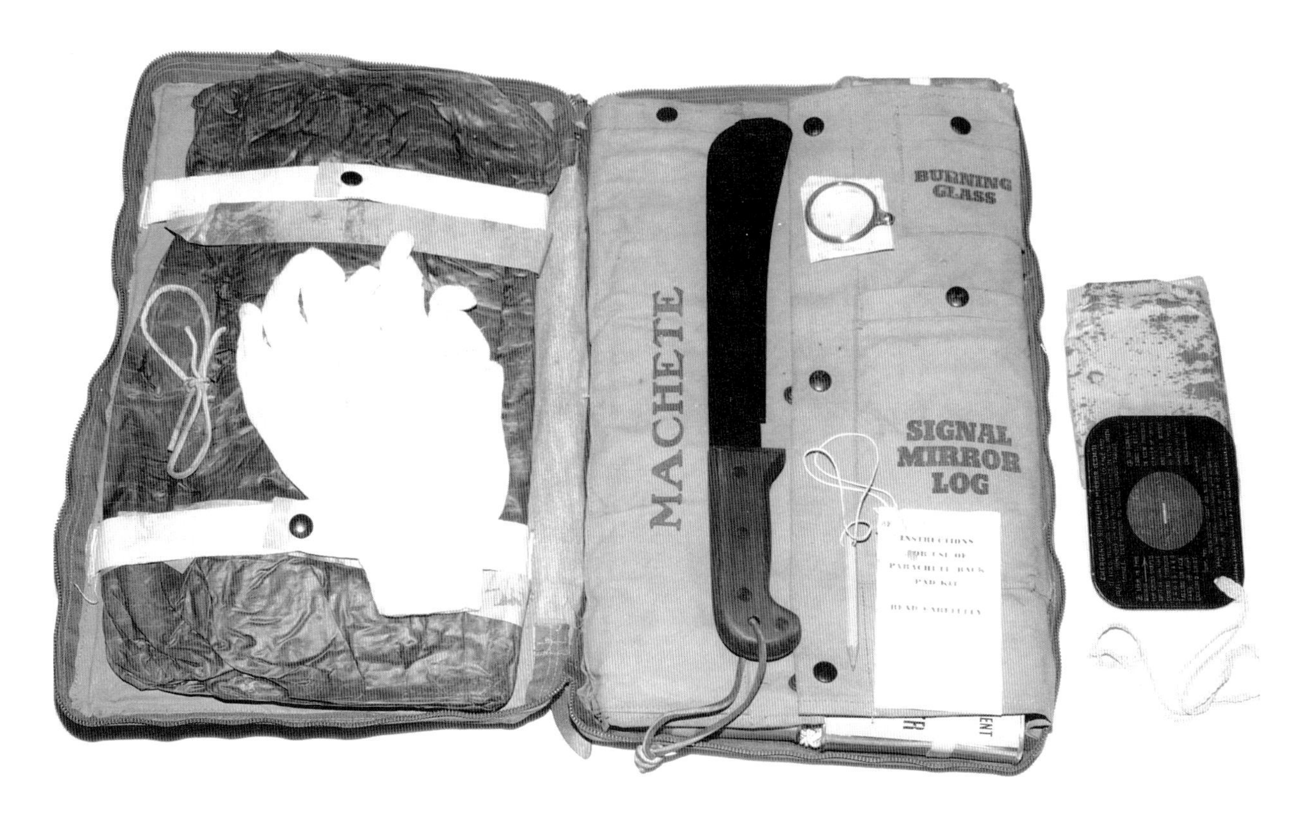

M-592 kit components, inside top cover and center compartment

M-592 BACK PAD KIT

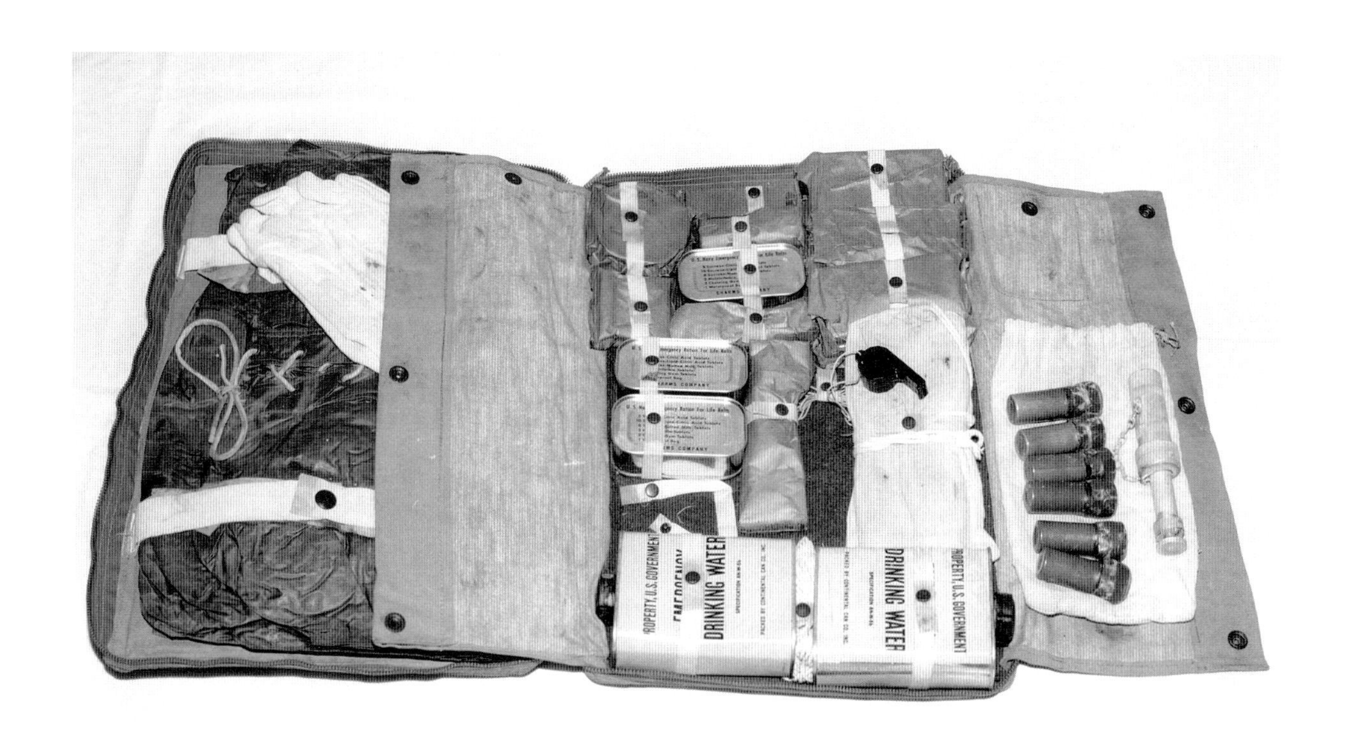

M-592 kit components, bottom compartment

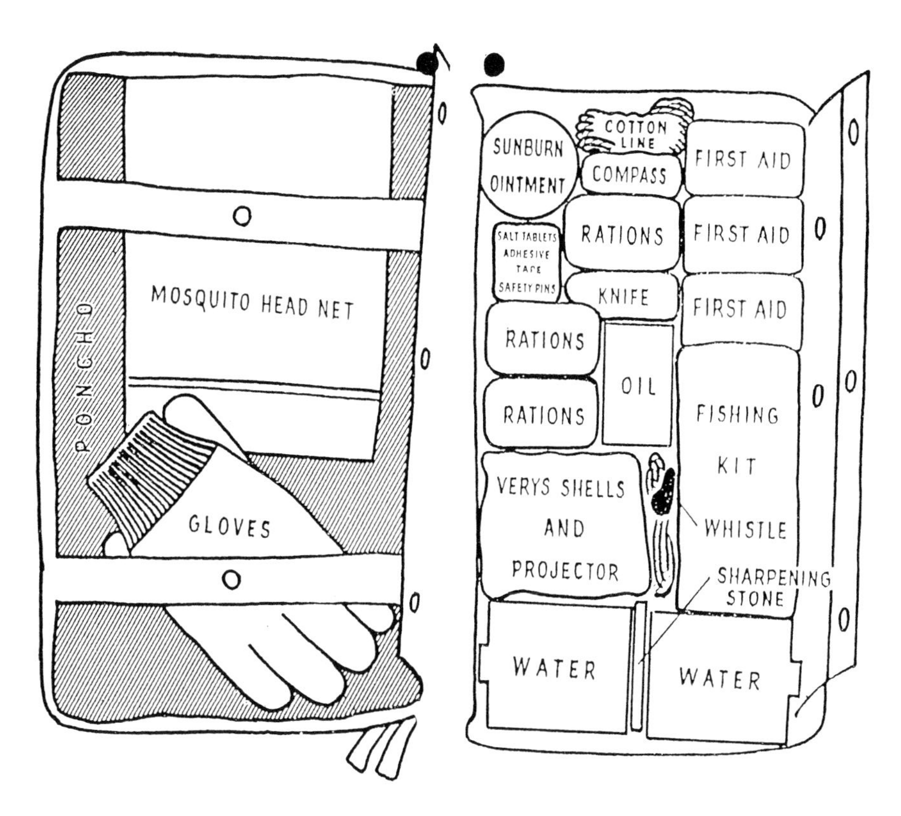

Location of items inside top cover and bottom compartment

M-592 BACK PAD KIT

U.S.N. M-592 Back Pad Kit, last variation

JUNGLE EMERGENCY PARACHUTE BACK PAD KIT

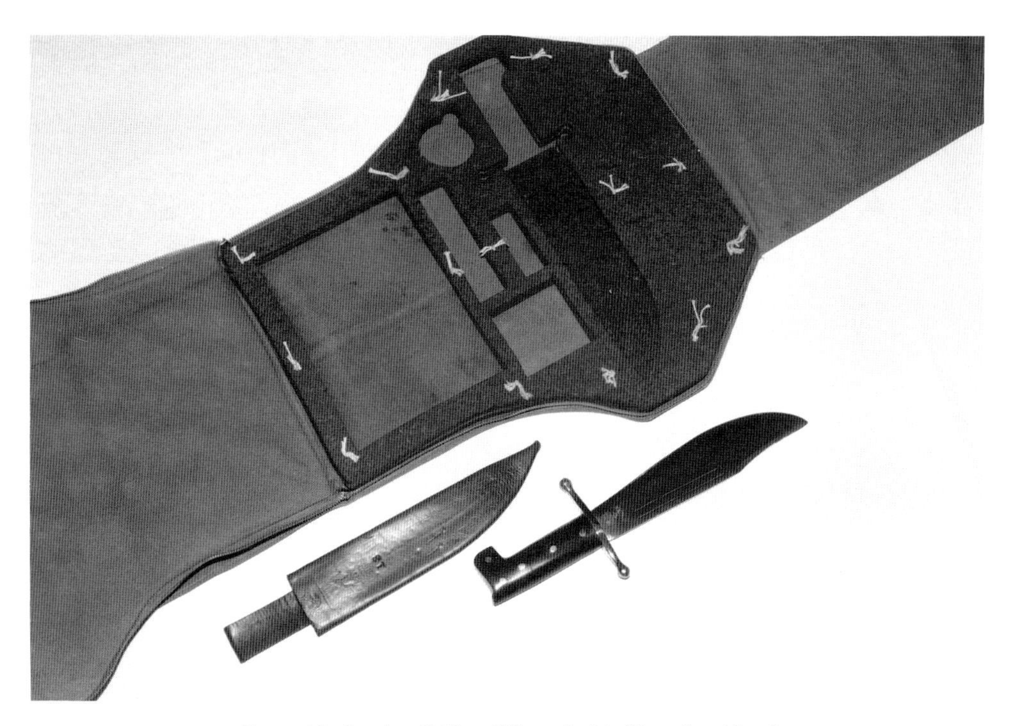

Empty kit showing Collins #18 survival knife and scabbard

The Jungle Emergency Parachute Back Pad Kit was one of the first survival kits issued to the A.A.F. and was designed to be attached to the parachute harness, taking the place of the standard back pad. This kit was the predecessor to the B-1 and B-2 back pad kits and was similar in size and shape. The 17" x 18 1/2" canvas container had a zipper closure on three sides and an inner felt/horsehair insert with cutouts for specific survival items. The only nomenclature was the part number, 38J5271, ink stamped on the outer casing.

Suggested components:

- Collins & Co. #18 'Bowie type' survival knife (14" overall length, 9 3/8" bright blade with brass cross guard and black composition handle)
- Brown leather tooled scabbard (number 13 stamped on front of scabbard)
- Compass
- Box of .45 caliber ball ammo
- Fishing kit (in small metal container)
- 3 Marples or Everdry match cases containing matches, 1/4 oz. bottles of iodine and quinine
- 2 Emergency Air Corp. Rations or 2 U.S. Army Field Rations 'D'

B-1 ALASKAN EMERGENCY PARACHUTE BACK PAD KIT

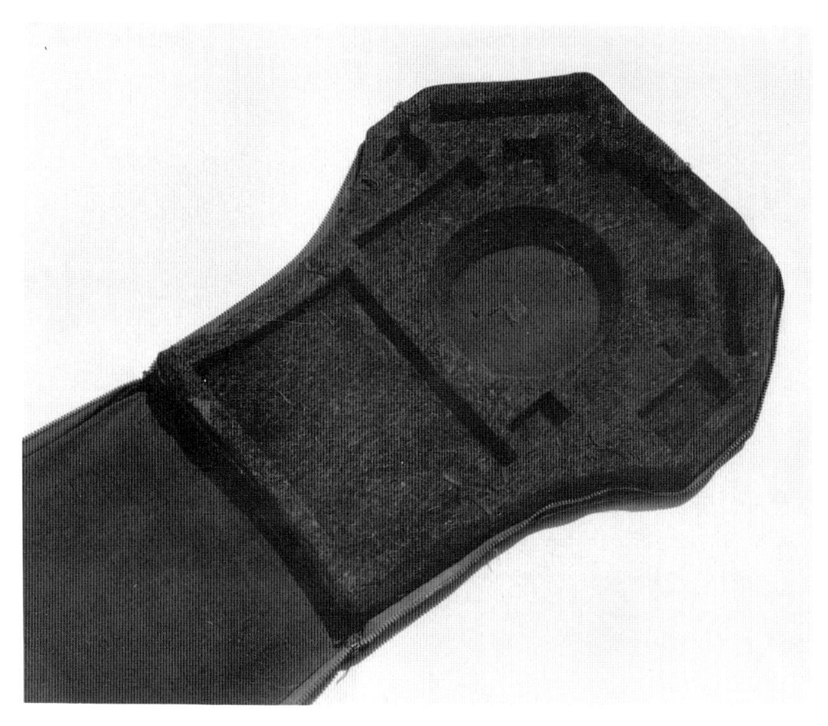

B-1 kit inner component arrangement

LABEL

TYPE B-1
SPECIFICATION NO. 40365
ORDER NO. 43-9344-AF
AIRCRAFT APPLIANCE CORP.
CHICAGO, ILLINOIS

The A.A.F. B-1 Alaskan Emergency Parachute Back Pad Kit was designed to be attached to the parachute harness (taking the place of the standard back pad) and provided flyers in northern regions with navigation equipment, gloves, emergency rations, cooking utensils and other supplies for emergency use in the event of a forced landing. The 17" X 18 1/2" canvas container had a zipper closure on three sides and an inner felt/horsehair insert with cutouts for specific survival items.

Suggested components for the B-1:

- -20 rounds of ammunition, .45 caliber ball
- -Pocket compass
- -Frying pan and detachable handle
- -Assorted fishing hooks, line, flies
- -D-2 gloves
- -Emergency rations (2 Emergency Air Corp Rations or 2 U.S. Army Field Rations 'D')
- -Insect repellent (2 oz. or 4 oz. bottle)
- -Waterproof match case and matches (Marples or Everdry type)
- -Miscellaneous first aid components (First aid components, i.e. iodine bottles, quinine, etc. were carried in empty Marples match cases for protection.)
- -Bouillon cubes
- -Pocket knife
- -Mosquito headnet

B-2 JUNGLE EMERGENCY BACK PAD KIT

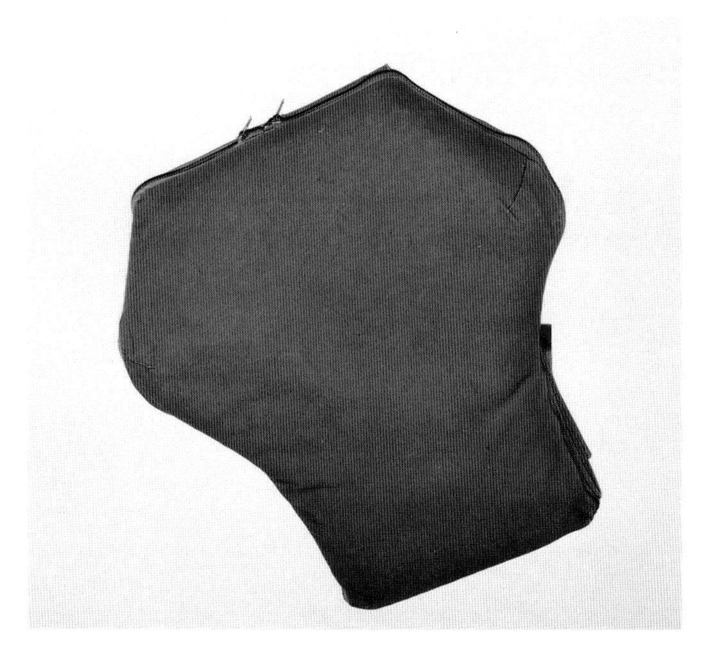

LABEL

TYPE B-2
SPEC. NO. 40401
ORDER NO. 42-16364-P
CABLE RAINCOAT CO.
PROPERTY
AIR FORCE, U.S.ARMY

Left: B-2 kit

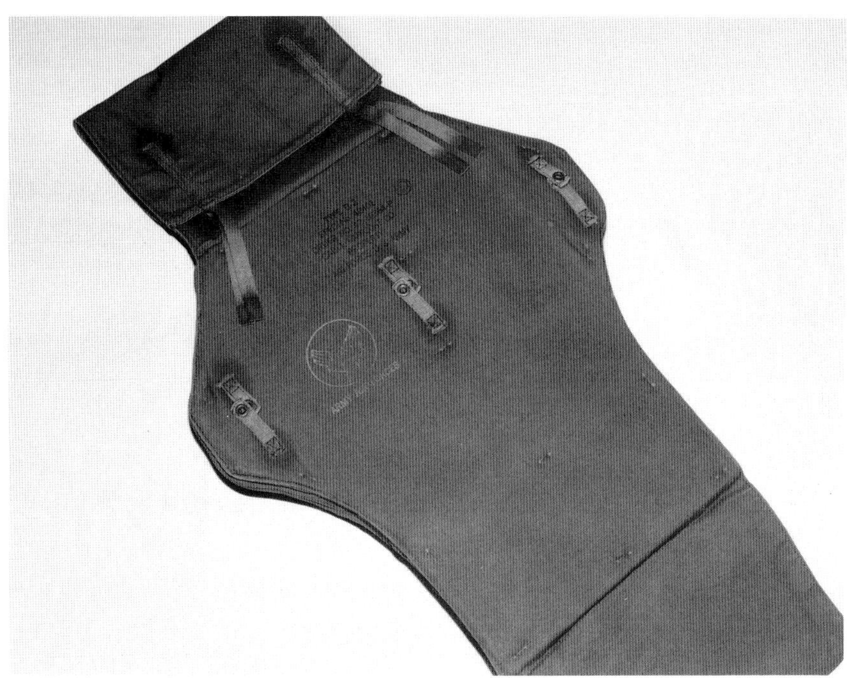

Right: B-2 kit (bottom)

The B-2 Jungle Emergency Back Pad Kit was produced in large quantities and was probably one of the most utilized A.A.F. back pad kits used during World War II. It was primarily designed for use in jungle areas and was attached directly to the parachute harness, taking the place of the standard back pad. This 17" x 18 1/2" olive drab container had an inner compartment made of felt/horsehair with cutouts for specific items. However, items could vary widely because the containers could be ordered separately and filled with locally purchased components. A 13" x 15" rectangular B-3 Jungle Emergency Seat Kit was also developed during World War II but not issued.

B-2 JUNGLE EMERGENCY BACK PAD KIT

Suggested components for the B-2 kit:

- -20 rounds, ball .45 caliber ammo
- -20 rounds, shot .45 caliber ammo
- -Pocket compass (U.S. stamped)
- -U.S. Army Field Ration 'D' (Three 4 oz. chocolate bars in 6 1/4" x 4 3/8" x 7/8" waxed paperboard container)
- -Sharpening stone (4" x 3/8" x 3/8")
- -Mosquito headnet
- -Waterproof container of matches (Marples, Everdry or clear plastic container)
- -10 inch folding blade machete
- -2 oz. or 4 oz. glass bottle of insect repellent
- -D-2 mechanics gloves
- -Fishing kit in metal container (3" x 1" x 3/4")
- -Pocket knife
- -Five minute flares, Fusee type (2)
- -Three pint water bladder
- -Bottle of halazone tablets
- -ESM-2 signal mirror
- -Folding sun goggle, type H-1
- -First aid kit:
  - Tourniquet
  - Aromatic ammonia inhalents
  - Burn ointment
  - Iodine
  - Compresses
  - Sulfanilamide

Right: WWII seat type parachute and seat cushion showing placement of the B-2 survival back pad kit

B-2 JUNGLE EMERGENCY BACK PAD KIT

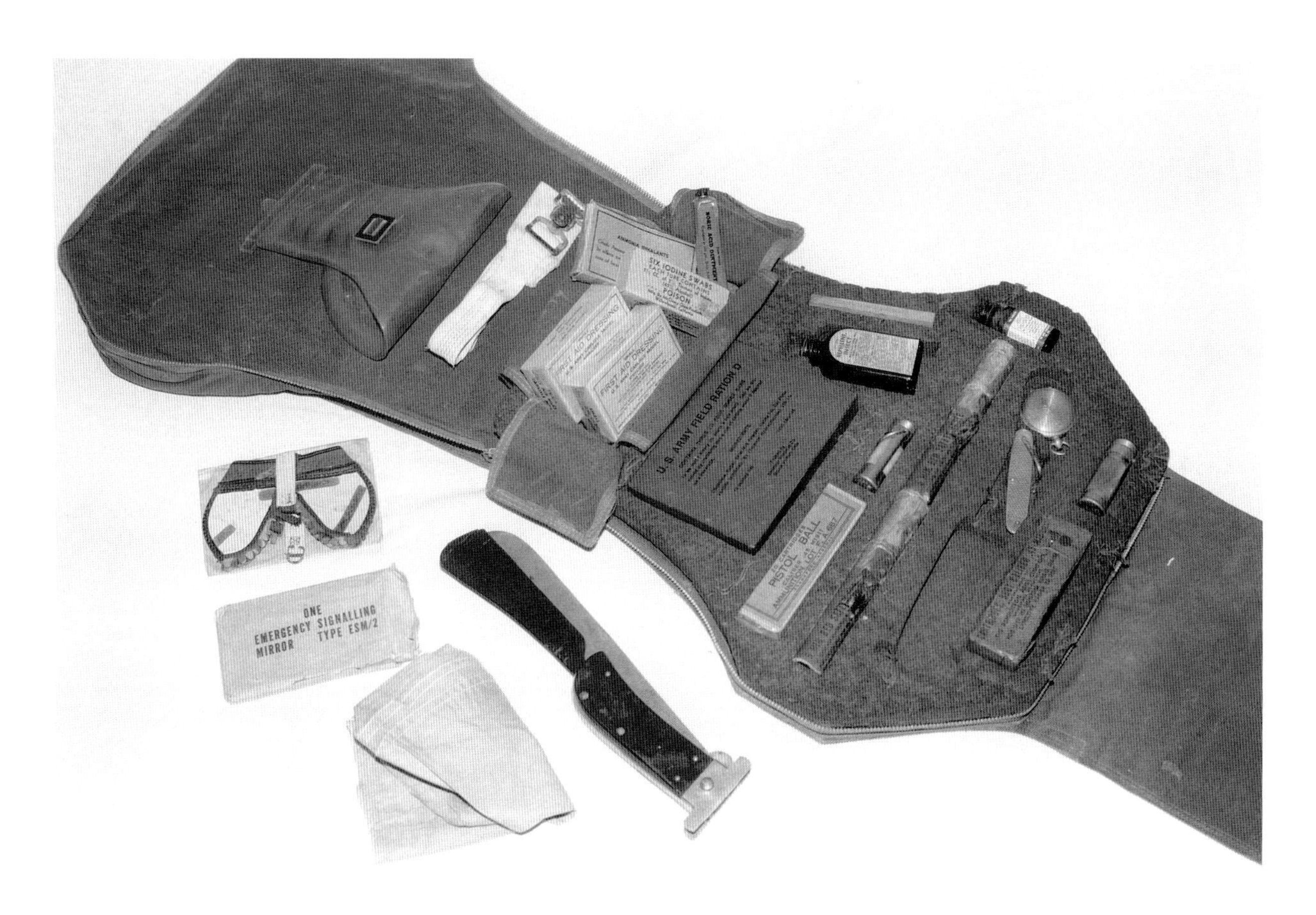

B-2 kit typical components (left to right, below kit): H-1 sun goggles, ESM-2 mirror, mosquito headnet, folding machete (left to right, in kit): 3 pint water bladder, tourniquet, first aid components, U.S. Army Field Ration 'D', box of .45 caliber ball ammo, match case, insect repellent, sharpening stone, 5 minute Fusee flares (2), halazone tablets, watch type compass, pocket knife, box of .45 caliber shot ammo, (Fishing kit is not shown.)

B-4 EMERGENCY PARACHUTE KIT

LABEL

TYPE B-4
SPECIFICATION NO. 40463
ORDER NO. 43-32225-AF
COMPANION BAGS INC.
PROPERTY
AIR FORCES, U.S. ARMY

Left: B-4 kit

The B-4 Emergency Parachute Kit was the last kit developed during World War II for the A.A.F. but used on a smaller basis than the B-2 kit. This 15" x 13" x 4" kit was designed for use in arctic, desert and tropical regions and was also made to be worn between the parachute harness and either the back or seat of the flyer.

The most unique item in the B-4 kit was the combination poncho/sleeping bag/shelter. This 60" x 78" yellow/olive drab down filled quilt was compressed into a container approximately 15" x 12" x 1 1/2" and was fit into the upper flap of the B-4 kit.

Right: Poncho/sleeping bag/shelter

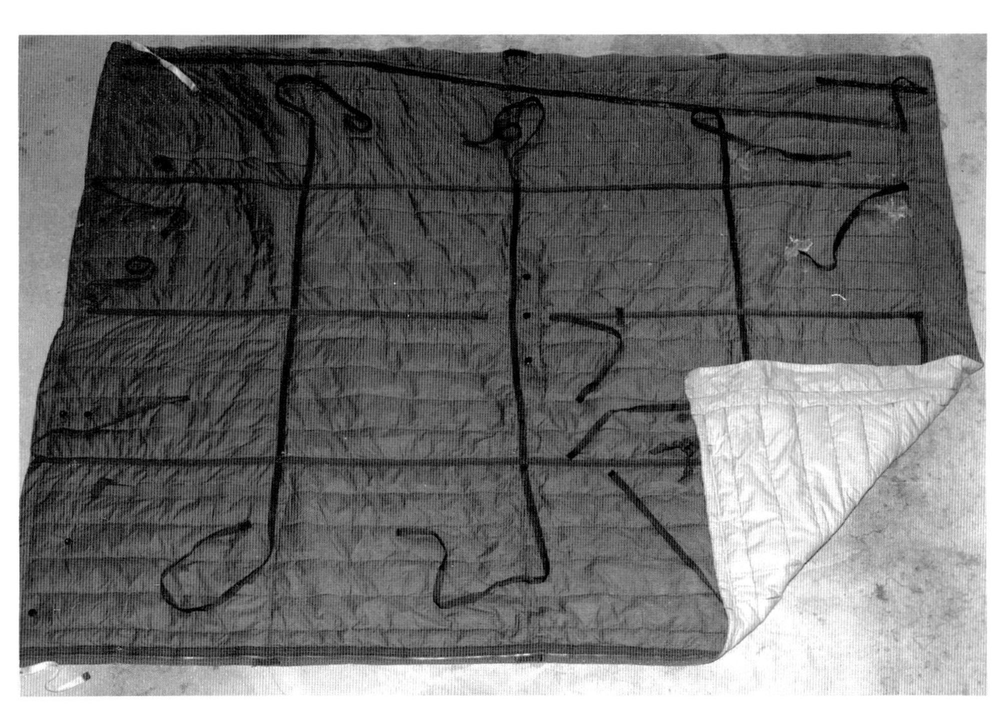

LABEL

FETH-R-PONCHO TYPE 2
PATENT PENDING
THOMAS QUILT FACTORIES
DENVER, COLORADO

B-4 EMERGENCY PARACHUTE KIT

Suggested components for the B-4 kit:

- -Parachute Ration (6 5/8" x 4 1/8" x 1 7/16")
- -Plastic water bladder, 3 pint
- -Mosquito headnet
- -ESM-2 signal mirror
- -Match case and matches (Marples type or clear plastic)
- -Pocket knife
- -Ten inch folding blade machete
- -Folding sun goggles, type H-1
- -D-2 mechanics gloves
- -Five minute flares, Fusee type (2)
- -Pocket compass
- -Ammo .45 caliber shot, 20 rounds
- -Frying pan and cover
- -Frying pan handle
- -Fishing kit
- -21W survival manual
- -First aid kit assembly (1" x 5" plastic circular container, inserted in frying pan)
- -Combination poncho/sleeping bag/shelter
- -First aid kit (frying pan insert) B-4 kit
  - Bandaids (6)
  - Sulfanilimide powder, 5 grams
  - Bar of soap
  - 2" compresses (2)
  - Needle and thread
  - Box of Benzedrine sulfate tabs (6)
  - Salt tabs (8)
  - Atabrine tabs (12)
  - Halazone tabs (30)
  - Tube of boric acid ointment
  - Iodine containers (6)
  - Tea tabs (3)

Right: Type B-4 first aid kit (frying pan insert)

B-4 EMERGENCY PARACHUTE KIT

B-4 kit typical components (left to right, in kit): survival manual - 21W and 3 pint water bladder on top of sealed poncho/ quilt, 5 minute Fusee flares (2), Parachute Ration, first aid kit, clear plastic match case, pocket knife, watch type compass, sharpening stone, folding machete (left to right, below kit): gloves, H-1 sun goggles, .45 caliber shot ammo, ESM-2 mirror, mosquito headnet (missing frying pan and handle, and fishing kit)

PK-2 PARARAFT KIT

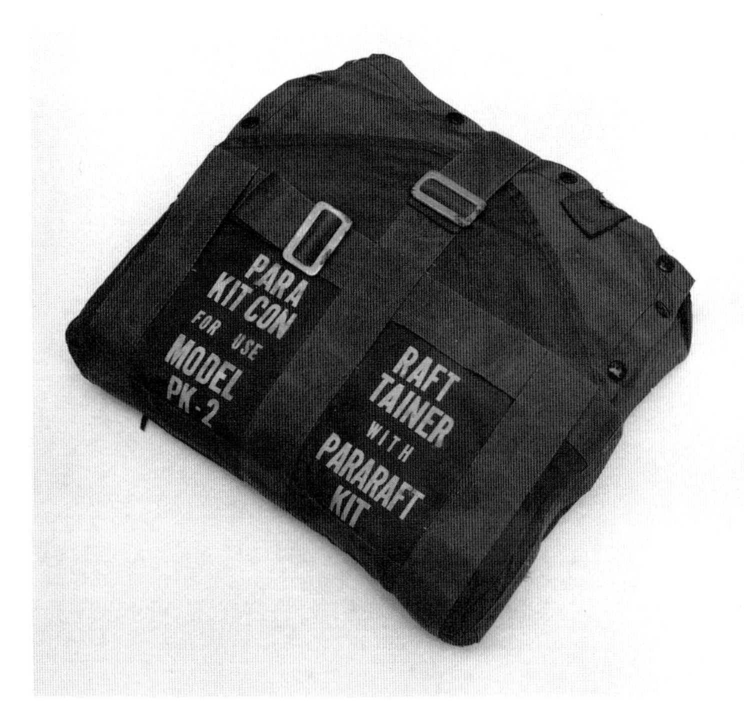

PK-2 kit outer case

LABEL

CARRYING CASE
PK-2 PARARAFT KIT
STOCK NO. R83K709970
CONTRACT NO. N383-31487A
NEW YORK RUBBER CORPORATION
U.S.PROPERTY

RAFT LABEL

PARARAFT-MODEL PK-2
SPECIFICATION MIL-K-8664(AER)
NEW YORK RUBBER CORPORATION
CONTRACT N383-31487A
DATE: JAN.1957

The Pk-2 Pararaft Kit was used by the U. S. Navy in the 1950's, 60's, and 70's and was carried on the parachute harness for emergency bailout over land and water in temperate climates. If a seat parachute was worn, the PK-2 was secured between the wearer's buttock and the parachute. The 15" x 13 1/2" x 3 1/2" kit consisted of two compartments: one containing the PK-2 raft and CO_2 inflation cylinder and the other compartment containing the survival equipment. The raft compartment flap was secured with snaps on three sides and the survival components were contained by a zippered flap. The PK-2 Pararaft is similar to the Army OV-1 'soft' type overwater survival seat kit except for the survival components.

PK-2 PARARAFT KIT

Components for the PK-2:

- -PK-2 raft and sea anchor (The raft was yellow with a spray shield colored grey on the ouside and bright purple on the inside.)
- -Water storage bag (5 qt.)
- -Nylon cord
- -Desalting kit
- -Sea dye marker
- -3" x 5" signal mirror
- -Poncho
- -MX-137/A radar reflector or AN-PRC-17 transceiver (early kits)
- -Sponge
- -Mk-13 Mod 0 smoke/flares
- -Sunburn ointment
- -Solar still

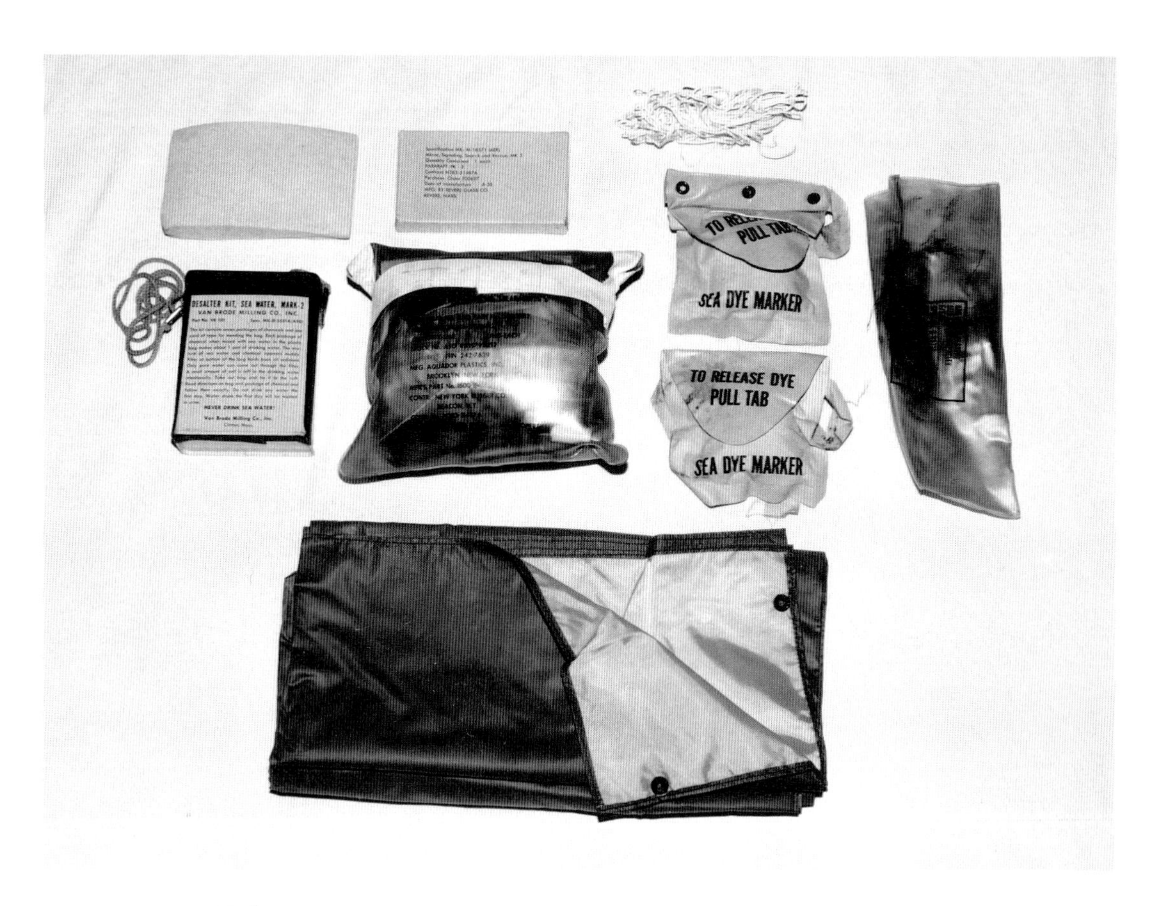

PK-2 components (left to right): sponge, desalter kit, mirror in box, solar still, paulin, sea dye marker, cotton line, vinylite container

CNU-1/P SUSTENANCE KIT CONTAINER

LABEL

SUSTENANCE KIT CONTAINER
BACK PAD PARACHUTE HARNESS
TYPE CNU-1/P
P/N 56E3848
S/N 4220-565-3275
LITE INDUSTRIES INC.
N383(MIS)66241A
MFD. 6/61
U.S.

Left: CNU-1/P container

The U.S.A.F. CNU-1/P Sustenance Kit Container was intended for use with back and seat style parachute harness assemblies and when installed replaced the conventional back pad cushion. The container, measuring 21 3/4" x 14" x 2", consists of an outer nylon casing with a slide fastener on three sides enclosing an inner 1 5/8" thick Ensolite spacer having cutouts or recesses for specific survival items. An inner cloth of ripstock canopy material helps hold the items in place. Harness keepers, directional dot fasteners and straps are provided on the outer casing for attaching the kit to the B-4 or B-5 back parachute and various seat style parachutes. When assembled, the kit becomes a fixed type in that items can not be varied. The back pad kit supplements a seat style survival kit (i.e. MD-1), or it may be used alone. If used alone the kit can provide a minimum amount of survival equipment needed to sustain an individual for four days where water is not a problem.

The CNU-1/P can be used in ejection seat aircraft provided seven inches of back depth can be provided (five inches for the parachute and two inches for the kit). Aircraft of the era applicable were the B-47, F-84, F-86, F-89, F-94 and some early B-52 and F-100.

Components for the CNU-1/P:

-First aid kit (carried in a clear plastic bag with a string tie and button closure)
-Button compass
-Lensatic compass (engineers type)
-Sun goggles, folding arctic in leatherette case
-Pocket knife
-ESM-2 or B-1 3" x 5" signal mirror
-Fishing kit in hinged plastic case
-Mk-13 Mod 0 smoke/flares (2)
-Matches and waterproof match case
-Razor and blades
-Heat tabs and folding stove
-Rations (SA or ST type) (2)
-Round sharpening stone 2 1/8" dia. x 3/4" (not in some kits)
-Wrist watch
-AFM 64-4 'Survival'

CNU-1/P SUSTENANCE KIT CONTAINER

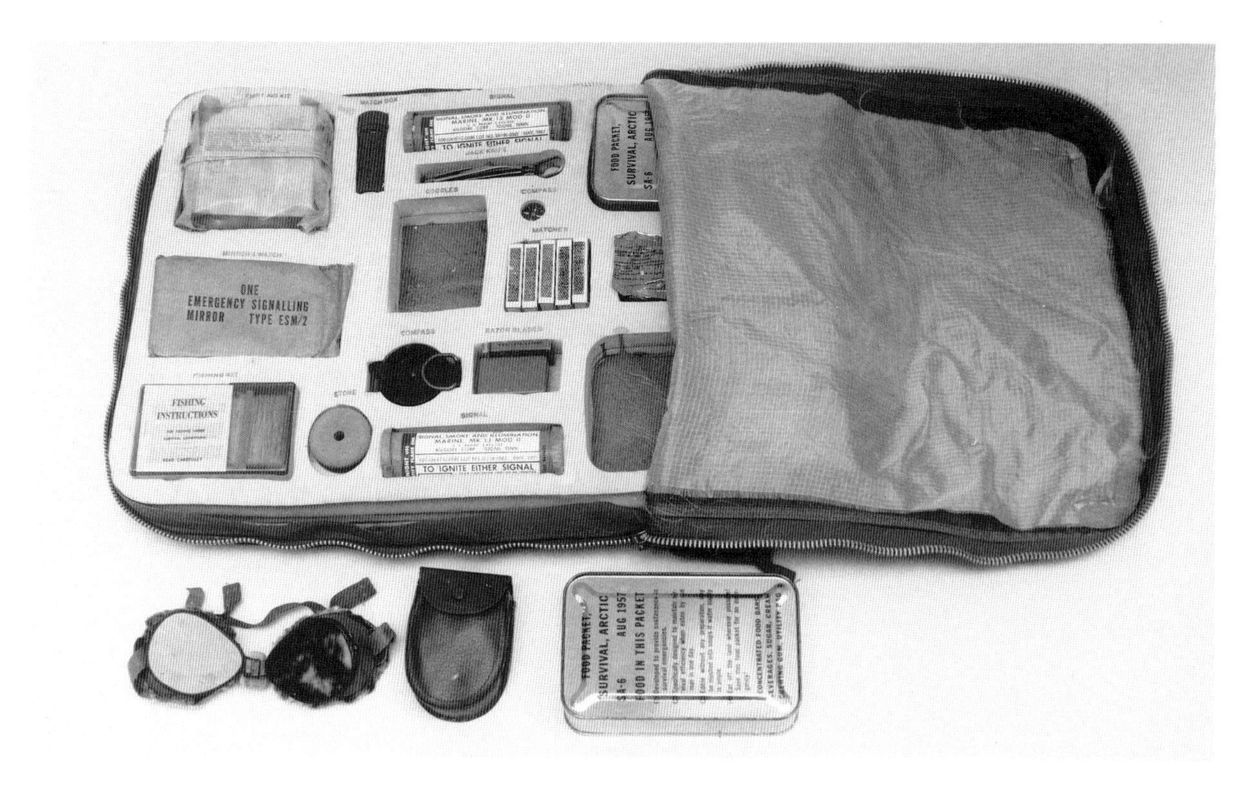

CNU-1/P components (left to right): first aid kit, ESM-2 mirrror, fishing kit, sharpening stone, engineers compass, Mk-13 Mod 0 smoke/flares, matches, razor blades, button compass, pocket knife, match case, trioxane fuel, SA ration (below case): sun goggles and case, SA ration

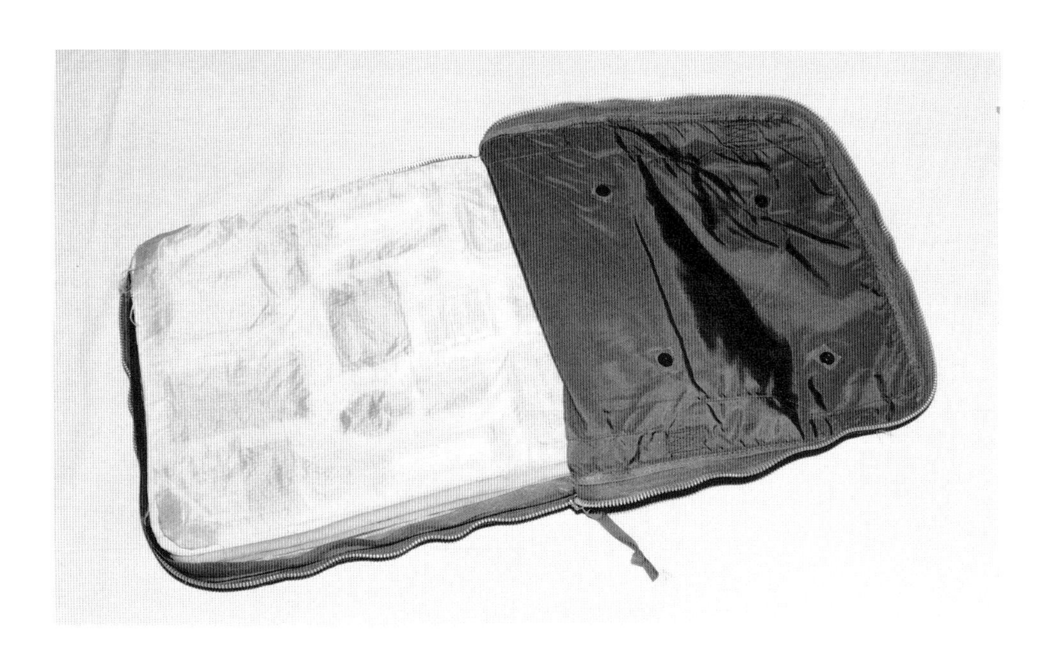

CNU-1/P showing ripstop nylon component retaining cover

ML-3 INDIVIDUAL LONG RANGE SURVIVAL KIT CONTAINER

ML-3 container (front)

LABEL

NAME OF ITEM: **CONTAINER ASSY.**
PART NO. 50J3944
ORDER NO. F09603-67-C-2990
MFRS. NAME: **THE CONRAD COMPANY**
PROPERTY OF U.S.AIR FORCE

ML-3 container (rear)

The ML-3 Individual Long Range Survival Kit Container (non-ejection seat) is a sage green nylon container approximately 15" x 15" x 10" in the packed configuration. Shoulder harneesses are proved for fashioning the container into a rucksack when ground travel is required and are stored inside the container until ready for use. Two snap pockets are located on either side of the outer container for storage of smaller items. The ML-3 could carry the M-4 or M-6 survival rifle and "Rifle Serial No." was printed on the top outside flap of the bag. The actual serial number was added after inclusion of the weapon. Because of the ML-3's large size, a wide variety of survival items could be carried. The container is not normally designed for bailout but may be attached to 'D' rings on back or chest style parachutes. Two ejector snaps are provided on the ML-3 for connection to the 'D' rings.

The ML-3 type container has been used on transport, cargo and early bombers and refueling aircraft since the 1950's and was stored so as to be readily accessible to the crew member. A similar container used in the 1950's was designated the Type E-1. This earlier kit contained a skillet assembly which the later ML-3 does not.

ML-3 INDIVIDUAL LONG RANGE SURVIVAL KIT CONTAINER

Mandatory components for the ML-3 container:

ITEM	STOCK NUMBER
-Mirror, MK-3	6350-299-6197
-Waterproof match box w/matches (3)	8465-265-4925
-Mk-13 Mod 0 signal (2)	1370-309-5027-L275
-Whistle	8465-254-8803
-Sleeping Bag SRU-15/P (or equal)	8465-753-3226
-First aid kit	6545-611-0978

Optional items are recommended for installation in this kit. This is to effect standardization within commands. However, if command mission requires the use of items other than indicated component variation shall be made.

Optional items in the ML-3 container:

ITEM	STOCK NUMBER
-Mukluk boot, twelve inches high, adjustable size	8430-275-5732
-Poncho	8405-290-0550
-Wool socks	8440-153-6717
-Hat and mosquito net	8415-269-0492
-Goggles, sun type MA1	8465-530-4083
-Food packet, survival, abandon aircraft	8970-753-6246
-Mitten, type N4B	8415-268-8312
-Gill net	4240-300-2138
-Self-locking snare, or 20 ft. brass wire (2)	4240-300-3294
-Long burning candle	6260-840-5578
-Saw, hand type MB-2	5110-570-6896
-Pocket knife	7340-162-2205
-Sharpening stone type VIII	5345-281-9485
-Water bag, size B	8465-634-4499
-Binocular	6650-938-6749
-Lensatic compass	6605-846-7618
-Insect repellent, stick form	Commercial
-Fire starter (2)	1370-219-8566-L621
-Heat tabs (1 box)	Commercial J.W. Speaker
-File, flat 6 inches	5110-234-6532
-Hood, winter	8415-543-7130
-Drawers, waffle weave	8415-782-3226
-Undershirt	8415-270-2012
-Survival tool kit SRU-18/P	4240-065-6713
*-Rifle, survival M-4, caliber 22 hornet	1005-575-0070
*-Cartridge, ball caliber 22 hornet	

* These items shall be installed at option of using command.

ML-3 INDIVIDUAL LONG RANGE SURVIVAL KIT CONTAINER

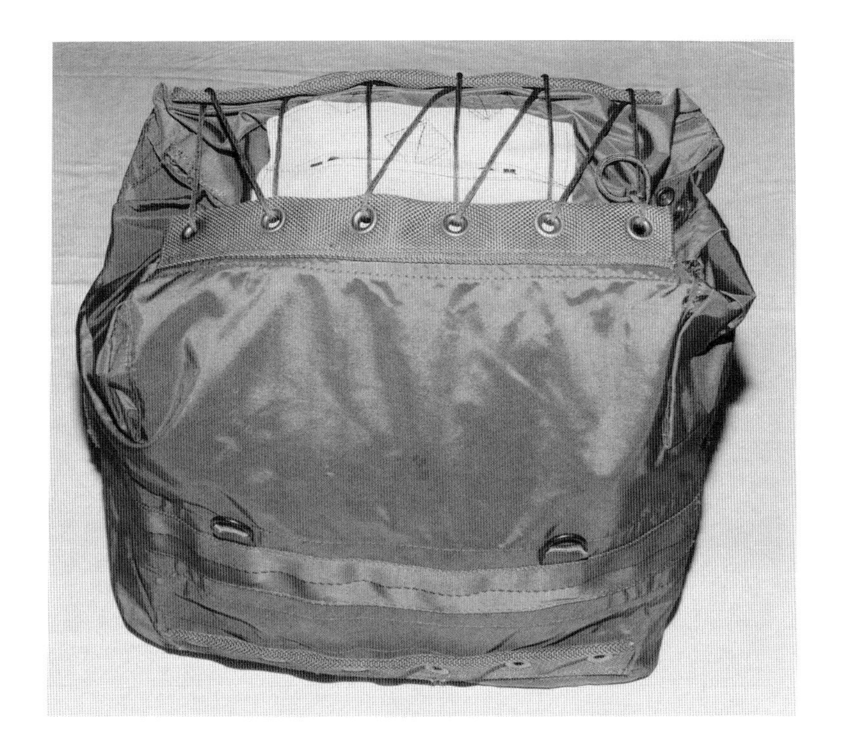

Left: ML-3 container (stowed configuration in aircraft)

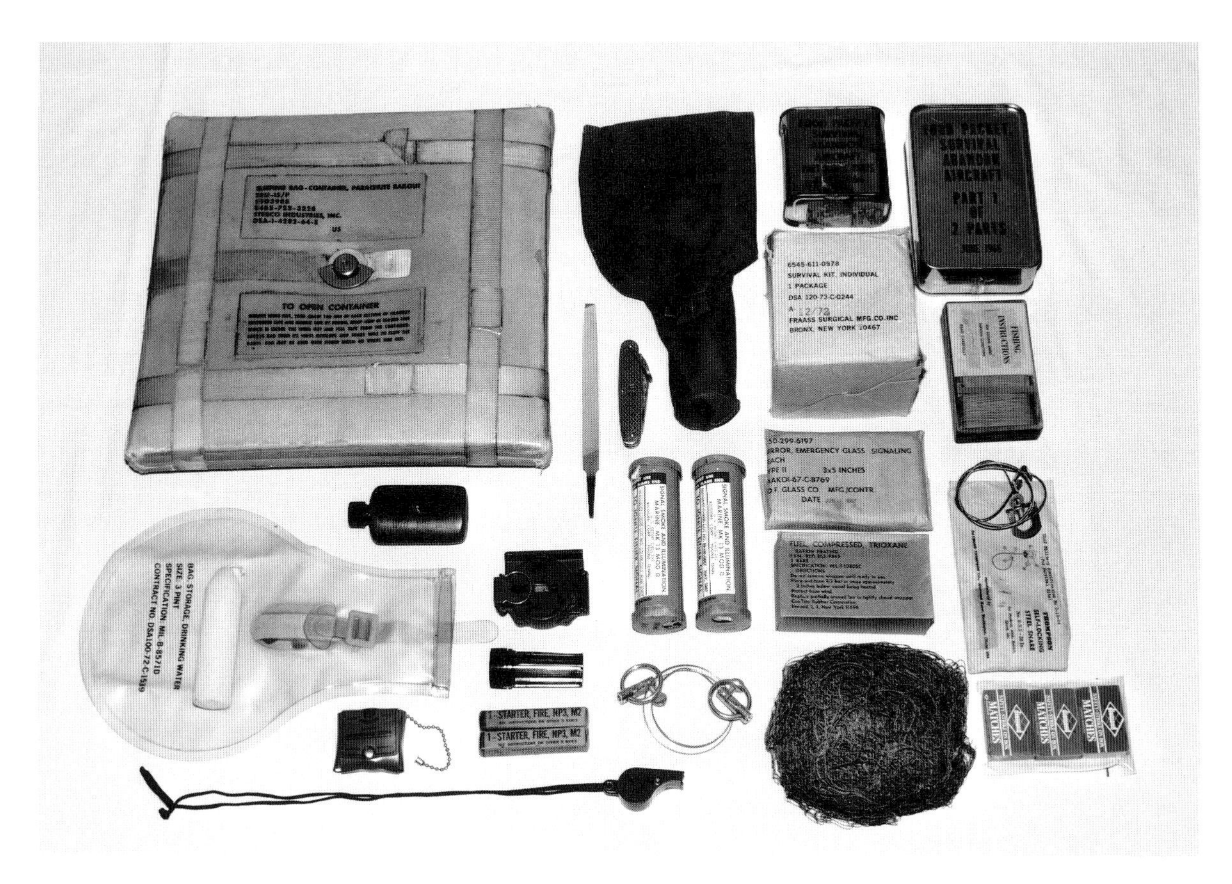

ML-3 container components (left to right): SRU-15/P sleeping bag, 3 pint water bladder, candle, whistle, sharpening stone, insect repellent, lensatic compass, match case, M-2 fire starters (2), finger saw, Mk-13 Mod 0 smoke/flares (2), 6" file, pocket knife, SRU-18/P survival tool, 2-part Abandon Aircraft ration, 2-part Individual Survival Kit (first aid kit), fishing kit, Mk-3 Type 2, 3" x 5" mirror in wrap, trioxane fuel , gill net, matches, snare

ML-3 INDIVIDUAL LONG RANGE SURVIVAL KIT CONTAINER

ML-3 components (continued) (left to right): poncho, AFM 64-5 survival manual, M-4 survival rifle and ammo, sun goggles and case, wool socks, N4B mittens, mosquito headnet, B-9B hat, mukluks

ML-4 RAFT KIT

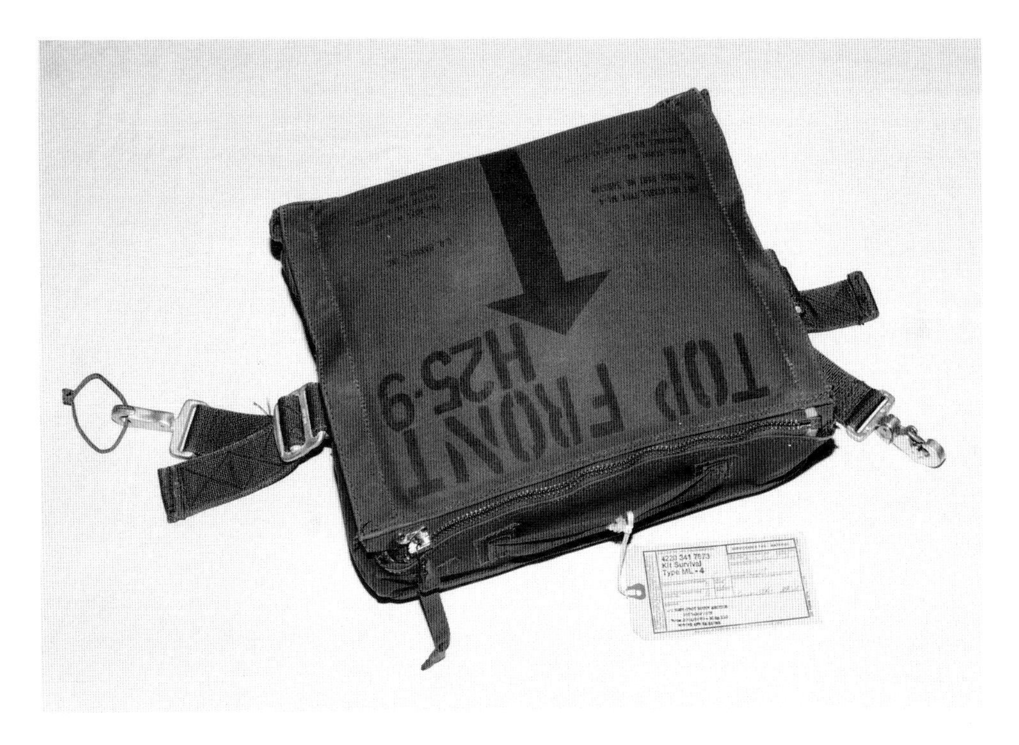

ML-4 kit

LABEL

RAFT INFLATABLE TYPE ML-4
AIR FORCE PART NO. 54D3749

MFRS SERIAL NO.
CONTRACT NO. DLA 700-82-C-0279
DATE OF MFR. 6-82

C.R. DANIELS, INC.

The ML-4 non-ejection seat style kit consists of an all fabric outer container, a non-waterproof inner container and a 25-foot drop lanyard for securing the raft and inner container to the wearer. A cushion is incorporated into the outer container's cover. The outer container, measuring 13" x 15" x 4 1/4", can be closed by means of snap fasteners along each side and a slide fastener along the front. The kit attaches to the parachute harness by one ejector snap assembly and one non-ejector snap. Because of limited storage space in the inner container, very few optional components in addition to the mandatory items can be utilized. A sleeping bag cannot be installed in this container.

Virtually any type of one-man raft can be used in the ML-4 and the raft is usually wrapped in a life raft retainer or 'diaper'. This cloth retainer is installed around the folded raft and will allow minor size and shape adjustments of the raft when being placed in either soft or rigid survival kits. Use of the 'diaper' will also eliminate the possibility of pinching the raft when closing the lids of containers.

ML-4 RAFT KIT

To inflate the raft, the crew member first opens the ejector snap on the left adjustable strap, disconnecting the left side of the kit from his parachute. He then grasps the slide fastener release lanyard on the right front edge of the container and pulls upward to remove the slide from the slide fastener and inflate the raft. The inflating raft will fully open the container and be deployed.

The ML-4 is used with back and seat style parachutes in cargo, transport and early B-50, and B-52 aircraft. It was listed as a substitute for the MB-2 and MD-1 seat kits in the F-86D and F-100.

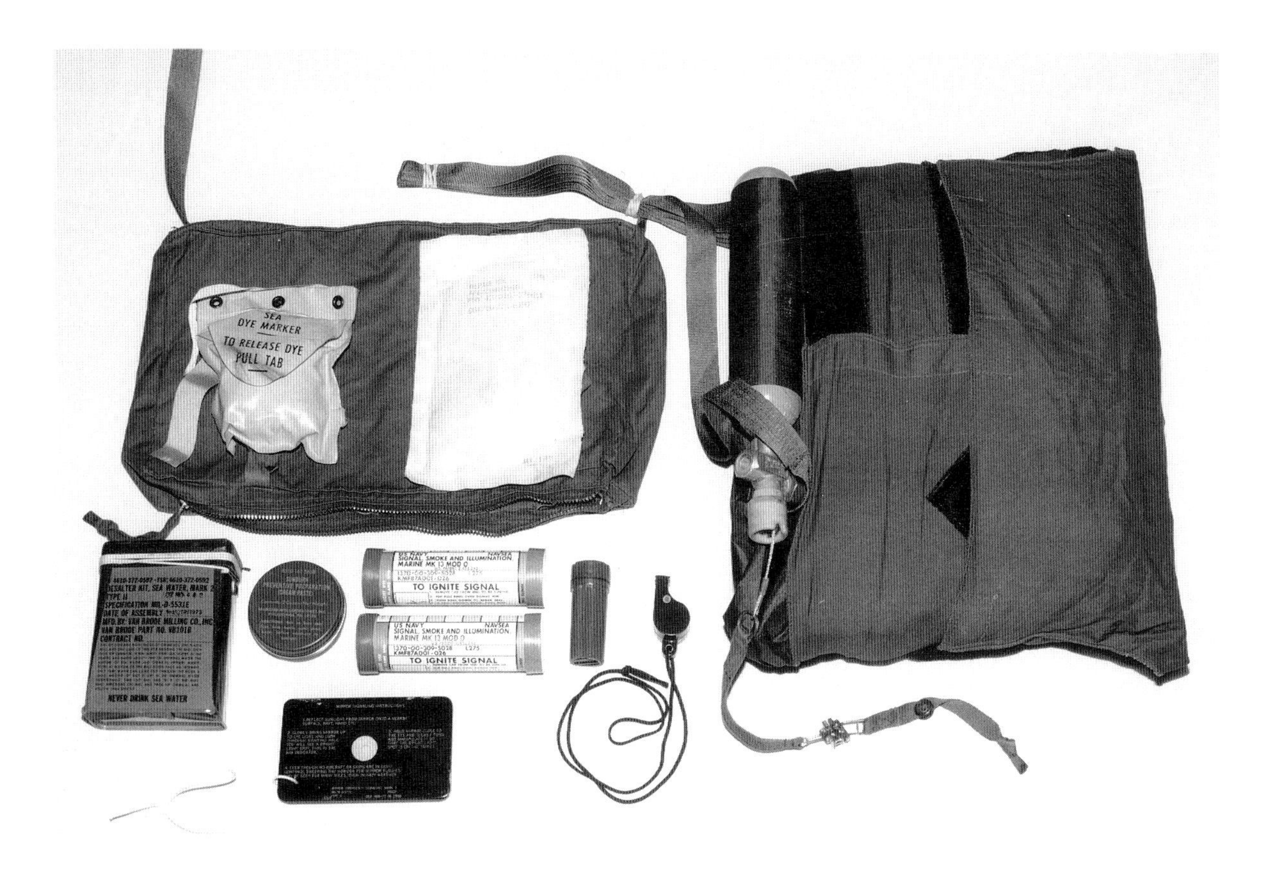

ML-4 kit components (left to right): inner container, sea dye marker, raft repair plugs in bag, desalter kit, sunburn ointment, Mk-13 Mod 0 smoke flares (2), Mk-3 3" x 5" mirror, match case, whistle, raft and CO2 cylinder in raft 'diaper'

<u>RAFT LABEL</u>

4220-01-003-6763LS
LIFE RAFT, INFLATABLE
LRU 16/P
SERIAL NO. 7381
RFD. PATTEN INC.
F41608-84-D-0003
MFD. NOV 1985

INDIVDUAL OVERWATER, HOT AND COLD CLIMATE SURVIVAL KITS

LABEL

US
SURVIVAL KIT
OVERWATER
NEW STAR TENT & AWNING CO.
MFD. 9/66
8465-082-2511

Left: Individual Overwater Kit

The Individual Overwater, Hot and Cold Climate Survival Kits have been in use since the early 60's and can be used on all Army and Air Force non-ejection seat aircraft, i.e. transports, helicopters, etc. The kits are stowed in readily accessible areas of the aircraft and can be attached to a parachute harness in case of bailout.

Except for their nomenclature, the three kits are almost identical outwardly. Each consisted of a 15" x 19" x 5 1/2" o.d. canvas container with a zipper closure on three sides and weighed about 28 to 29 pounds. A carrying handle and external stowage pocket (for a radio, extra water, etc.) are also attached. Some have yellow nylon adjustable back harnesses sewn to one side to facilitate carrying as a back pad once on the ground. Each kit has a smaller inner canvas zippered container with a nylon web attached to two parachute ejector snaps that route thru an opening on each side of the outer container.

A 14 1/4" x 13 1/2" x 2 1/2" aluminum 'frying pan' forms the basis of each kit and is inserted into the inner container. This pan can be used for cooking, collecting water, personal hygiene, etc. and has a small hole in each corner to aid in removal/placement from a fire by inserting a small stick or piece of wire.

Three types of first aid kits were used. The earlier type contained the medical components in a clear vinyl bag and used a string and button closure. This was followed by an o.d. green canvas container marked 'FIRST AID KIT, AVIATORS CAMOUFLAGED'. The last and current first aid kit uses a nylon outer bag with a plastic case insert. The last two kits could be carried from a belt.

All three kits shared certain similar items; insect headnet, fishing kit, five quart water bladder with carrying strap, pocket knife, matches and match case, Mk-13 Mod 0 smoke/flares, MC-1 compass, trioxane fuel, general purpose food rations and 3" x 5" Mk-3 signal mirror. The remaining items are unique to the particular seat kit.

The Individual Overwater Survival Kit contains either an LR-1, LRU-3/P, or LPU-16/P one-man life raft. These rafts were inflated by a CO2 cylinder that is manually activated by a pull on a lanyard attached to the outer bag zipper. Pulling the lanyard would remove the zipper keeper, inflating the raft and opening the cover on the container. The raft was atttached to the outer container by a twenty-five foot drop line.

The overwater kit contained a reversible colored hat; green on one side, orange on the other; two blue wooden paddles with reflective tape on one side; sponge; bailer; sea dye markers; raft repair kit; and three Mk-2 desalter kits.

INDIVIDUAL OVERWATER, HOT AND COLD CLIMATE SURVIVAL KITS

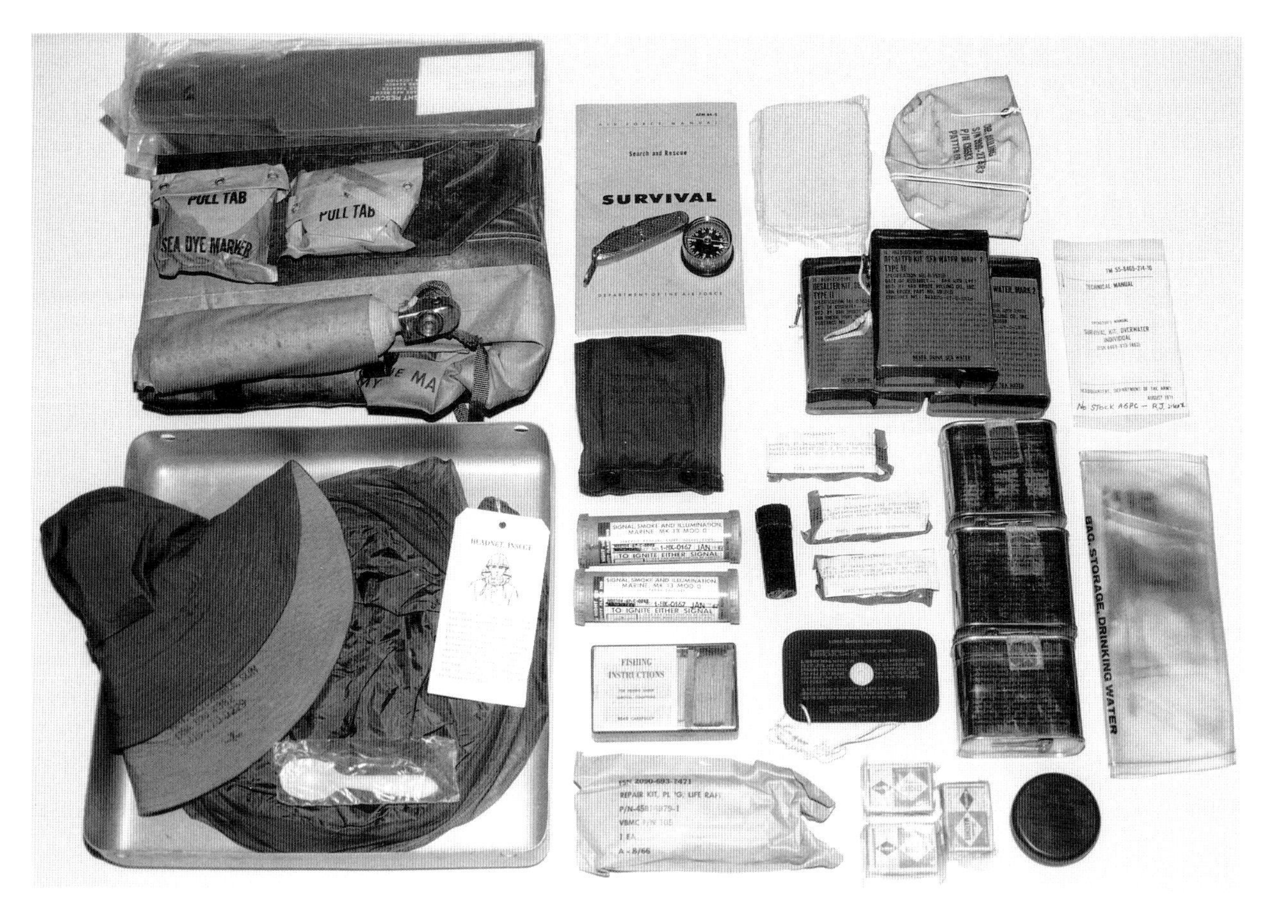

Individual Overwater Kit components (left to right): frying pan, reversible hat, spoon, mosquito headnet, LRU-3/P raft, paddles, sea dye marker, survival manual, pocket knife, MC-1 compass, Aviators Camouflaged First Aid Kit, Mk-13 Mod 0 smoke/flares (2), fishing kit, raft repair plugs in bag, matches, Mk-3 3" x 5" mirror, match case, trioxane fuel (3), desalter kits (3), sponge, bailing bucket, operators manual, rations (3), 5 quart water bladder, sunburn ointment

RAFT LABEL

LIFE RAFT, INFLATABLE, ONE MAN
TYPE-LRU-3/P U.S. ARMY
MIL-L-86648(WP)
RUBBER FABRICATORS, INC.
CONT. NO. DAAKO1-67-C-0916
DATE MFG. DEC. 1967
SERIAL NO.

The Individual Hot Climate Survival Kit had a multi-purpose ten-inch bladed 'machete' that incorporated an axe edge, sickle blade and trenching edge. The tool had a canvas sheath that could be hung from a belt and two inside pockets that held a round sharpening stone, glass burning lens and survival instructions. The blade is stamped or labeled 'FRANK & WARREN, INC.', 'SURVIVAL AX TYPE IV', MIL-S-8642C.

A 67 1/2" x 84" (size 77) paulin is in the hot climate kit. This paulin, tarpaulin in civilian terms, is yellow on one side and blue grey on the other and can be used as a shelter, to collect rain water or as a signal panel. The blue-grey side has instructional data for folding the paulin to form international rescue signals. This kit also contains twelve cans of drinking water, whistle, 20 feet of brass snare wire and a reversible sun hat.

INDIVIDUAL OVERWATER, HOT AND COLD CLIMATE SURVIVAL KITS

Individual Hot Climate Kit

LABEL

US
SURVIVAL KIT
HOT CLIMATE
NEW STAR TENT & AWNING CO.
MFD. 9/66
8465-082-2513

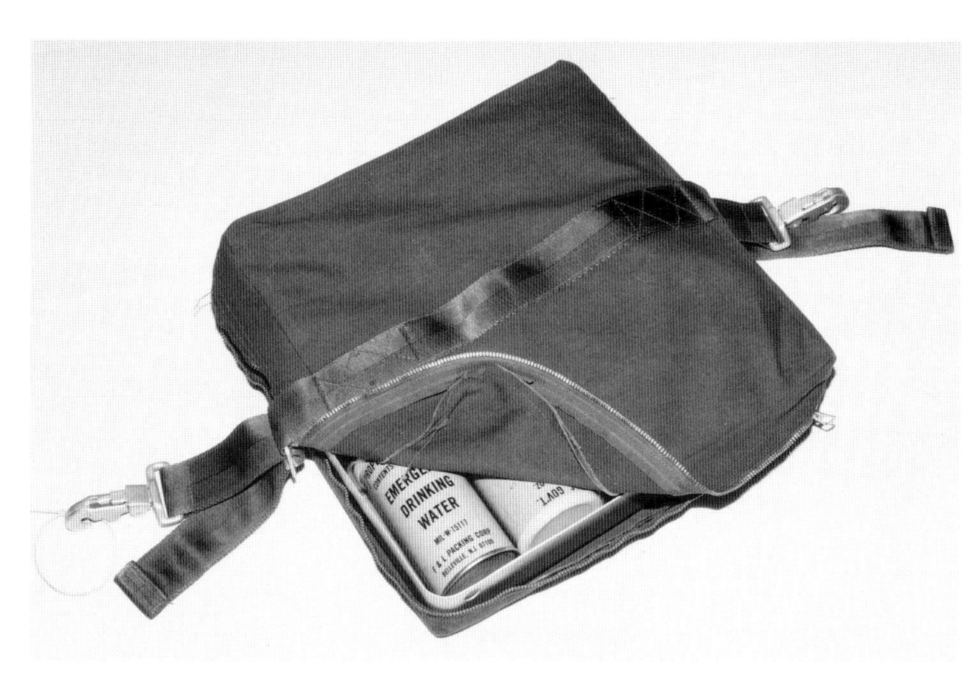

Individual Hot Climate inner container

INDIVIDUAL OVERWATER, HOT AND COLD CLIMATE SURVIVAL KITS

Individual Hot Climate Kit components (left to right): paulin, operators manual, pocket knife, frying pan, reversible hat, survival manual, spoon, mosquito headnet, rations (6), matches, Mk-3 3" x 5" mirror, snare wire, MC-1 compass Mk-13 Mod 0 smoke flares (2), fishing kit, Type IV survival tool, cans of water (12), sunburn ointment, first aid kit, 5 quart water bladder, whistle, match case, trioxane fuel

The Individual Cold Climate Survival Kit had some unique items geared to the colder climates. A down filled, mummy type sleeping bag (designated SRU-15/P) is compressed in a fiberglass container measuring 13" x 13 1/4" x 2". Once the sleeping bag has been removed it can not be returned to the container and must be carried separately. A standard cold climate or artic type sleeping bag can be found in this kit.

A type A-2 saw/knife/shovel assembly is a multi-part tool with a removable red handle that can be attached to either a 14" x 13" shovel or a 14 1/4" long saw/knife. The shovel can be used to shovel snow or make snow blocks for a shelter and the saw/knife can be used for cutting small trees or limbs, cutting holes in the ice for fishing or when dressing game.

A dark green poncho is in the cold climate kit. The poncho has a drawstring hood and snap fasteners around its perimeter to allow its use with other ponchos to form a shelter or to just form a closed garment when worn by an individual. Five wax candles are carried and each can burn up to eight hours. Candles in the newer cold climate kits are supplied in their own individual aluminum container.

INDIVIDUAL OVERWATER, HOT AND COLD CLIMATE SURVIVAL KITS

Individual Cold Climate Kit

LABEL

US
SURVIVAL KIT
COLD CLIMATE
KINGS POINT MFG. CO., INC.
DATE OF MFG. JUL 82
1680-00-082-2512

LABEL

U.S.
SURVIVAL KIT OVER-WATER
8465-782-3032
Mfr: OWEN MILLS, INC.
DSA 100-69-C-1864

Individual Cold Climate Kit using a modified Overwater outer container

INDIVIDUAL OVERWATER, HOT AND COLD CLIMATE SURVIVAL KITS

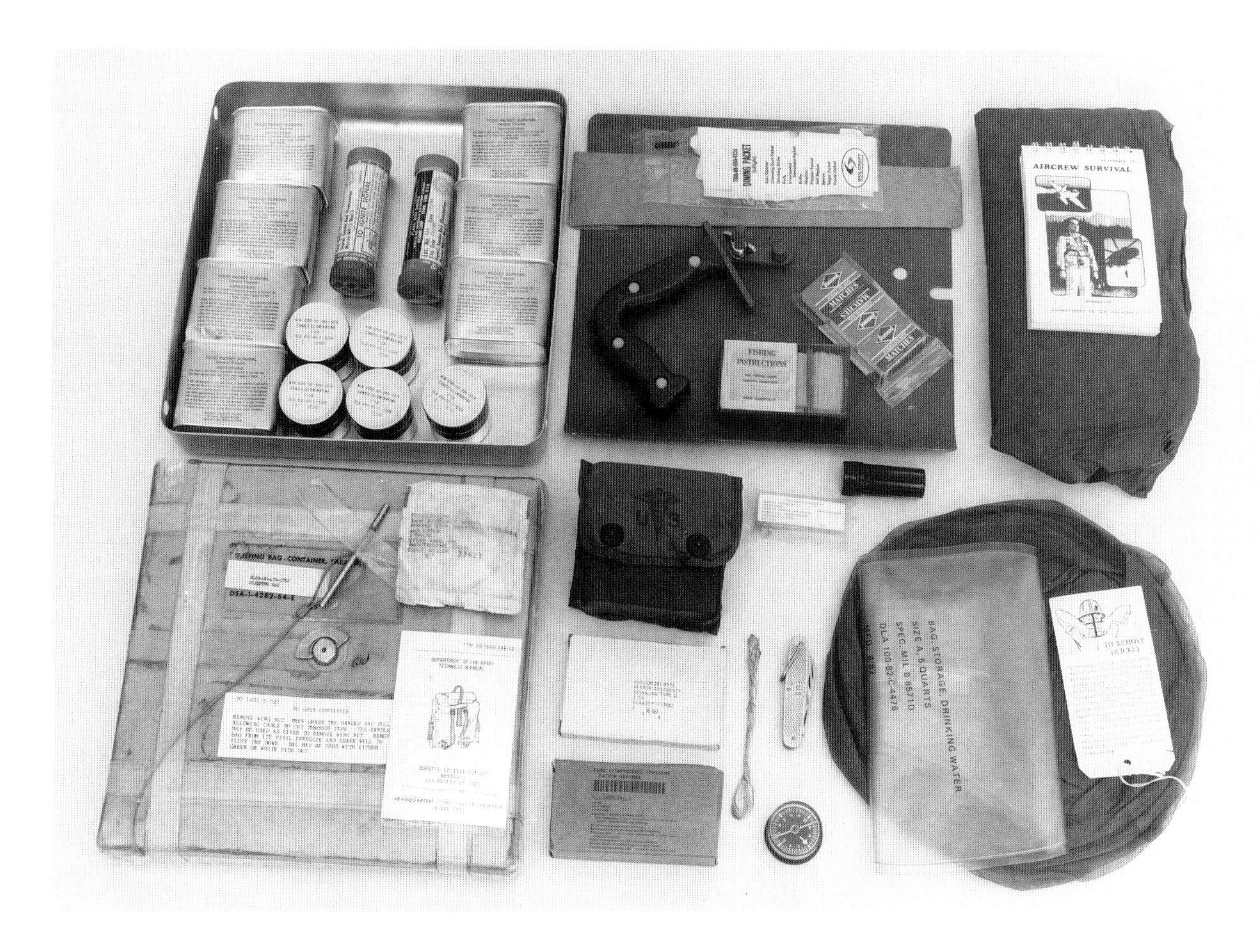

Individual Cold Climate Kit components (left to right): SRU-15/P sleeping bag, operators manual, finger saw in bag, frying pan, rations (7), Mk-13 Mod 0 smoke/flares (2), 5 cans containing candles, A-2 tool showing handle, shovel blade, saw/blade in paperwrap, fishing kit, matches, dining packet containing spoon, spices, etc., first aid kit, Mk-3 mirror in box, trioxane fuel, snare wire, pocket knife, MC-1 compass, match case, magnesium fire starter, poncho, survival manual, mosquito headnet, 5 quart water bladder

Gill nets, magnesium fire starters and combat casualty blankets are being issued in more recent overwater, hot and cold climate kits. The gill net are used for fishing, carrying equipment/collected food, hanging cache and as a hammock or stretcher. The combat casualty blankets are used to provide warmth and protection against the elements and as a signaling aid for rescue purposes. The blankets consist of two types. The Type I heavyweight is 84" inches long by 56 inches wide. The Type II, lightweight is 96 inches long by 56 inches wide. One blanket is olive drab/silver, the other is orange/silver and made of aluminized plastic.

Assembly sheet drawings for each kit are included in the appendix.

EJECTION SEAT SURVIVAL KITS

U.S.A.F. F4 Phantom II Martin-Baker Mk-H7 ejection seat
using a CNU-111/P seat survival kit

The majority of ejection seat survival kit containers are fabricated of fiberglass, and are normally used with back style parachutes only. These 'hard'kits are designed to fit numerous fighter and bomber aircraft and many examples exist. Early ejection seats employed containers such as the MD-1 that used a fiberglass or magnesium seat pan and a nylon expandable lower section. The newest Air Force kits, such as those used in the ACES II ejection seat, are a departure from the 'hard' containers. A 'soft' container attaches to the crewmember and is installed into the seat pan and covered by a forward hinged lid. After ejection and upon seat separation, the hinged lid opens and the survival kit stays with the individual.

Some kits contain, in addition to the liferaft, survival equipment, and oxygen systems in case of high altitude ejection. The oxygen systems consist of a pressure demand regulator, emergency oxygen cylinder, oxygen pressure gauge, hoses, reducers and cables to facilitate delivery of oxygen to the crewmember from the aircraft or the emergency supply as required. The oxygen system would be contained in the back parachute or attached directly to the ejection seat if not part of the survival seat kit.

Although many variations exist, all seat kits consist of a main compartment, which houses the inner waterproof container with survival components; one-man raft and a 25 foot drop lanyard. The inner waterproof container can incorporate a set of straps to facilitate its use as a rucksack/backpack for ground travel. A rear compartment, if part of the container, can contain additional survival components, radio, etc. or the components of an oxygen system. Some kits contain the oxygen system in the lid of the main compartment when a rear compartment is not part of the design. Seat kit survival components can vary depending upon whether or not a survival vest is worn.

The outer container has quick disconnect hardware for securing the kit to the parachute harness and a yellow emergency handle for release of the kit after parachute deployment. A foam cushion is attached to the lid by snaps and Velcro tape. Radio beacons (PLB'S) are usually part of the kit and are secured internally with Velcro tape. After kit deployment and lid separation the PLB is automatically initiated and continues to transmit until the crewmember turns the unit off or until the battery discharges.

Because of the many variations of 'hard' type ejection seat survival kits that are used only several examples have been included. These kits are representative of the majority of kits employed.

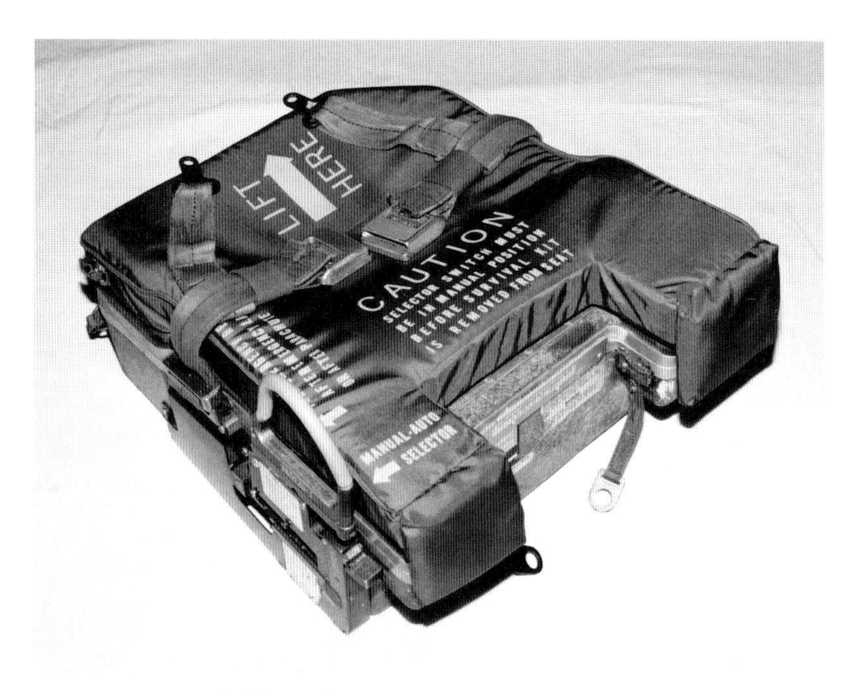

U.S.A.F. F4 Phantom II CNU-111/P seat survival kit
(A URT33B beacon is used in this kit.)

LABEL

CONTAINER, SURVIVAL KIT
TYPE CNU-111/P
CONTRACT NO. F33657-68-C-1024
SERIAL NO.
FSN 1660-104-3261-TP
H. KOCH & SONS, INC.
CORTE MADERA, CALIFORNIA
PART NO. 140000-100
PATENT NO. 3,107,370

EJECTION SEATS SURVIVAL KITS

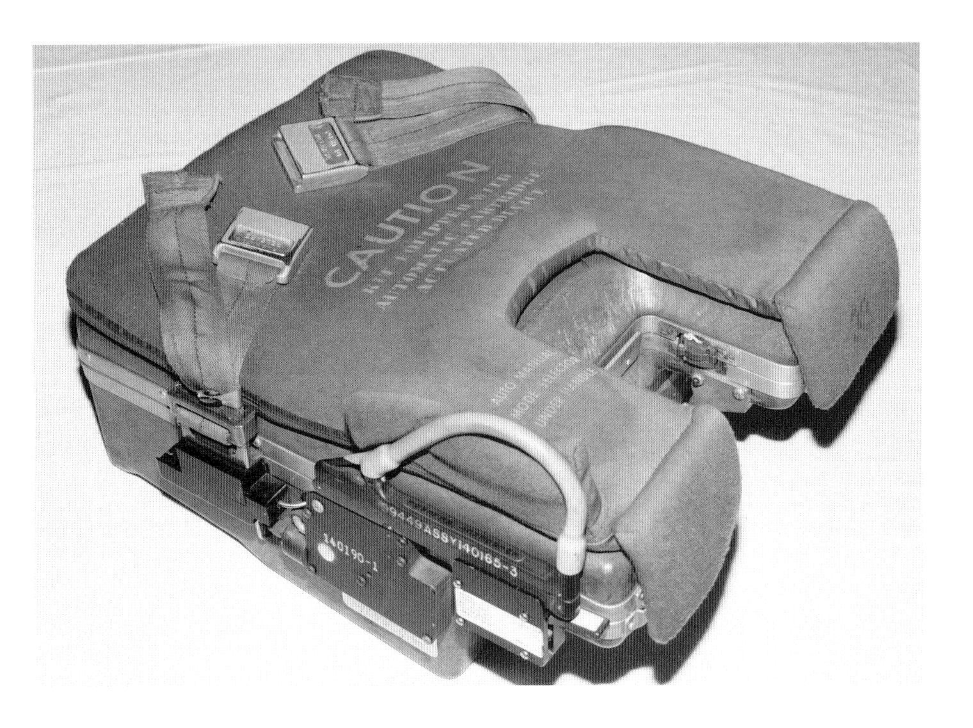

Left: U.S.A.F. A7D Corsair ejection seat survival kit used in the Escapac IC2 ejection seat

Right: URT-33B radio beacon location in Escapac IC2 seat kit

LABEL

CONTAINER, SURVIVAL KIT
WITH ACTUATOR
CONTR. N00019-70C-0497
V.A.D. PART NO. 216-21032-8
FSN
H. KOCH & SONS, INC
CORTE MADERA, CALIFORNIA
PART NO. 140000-135
PATENT NO. 3,107,370

EJECTION SEAT SURVIVAL KITS

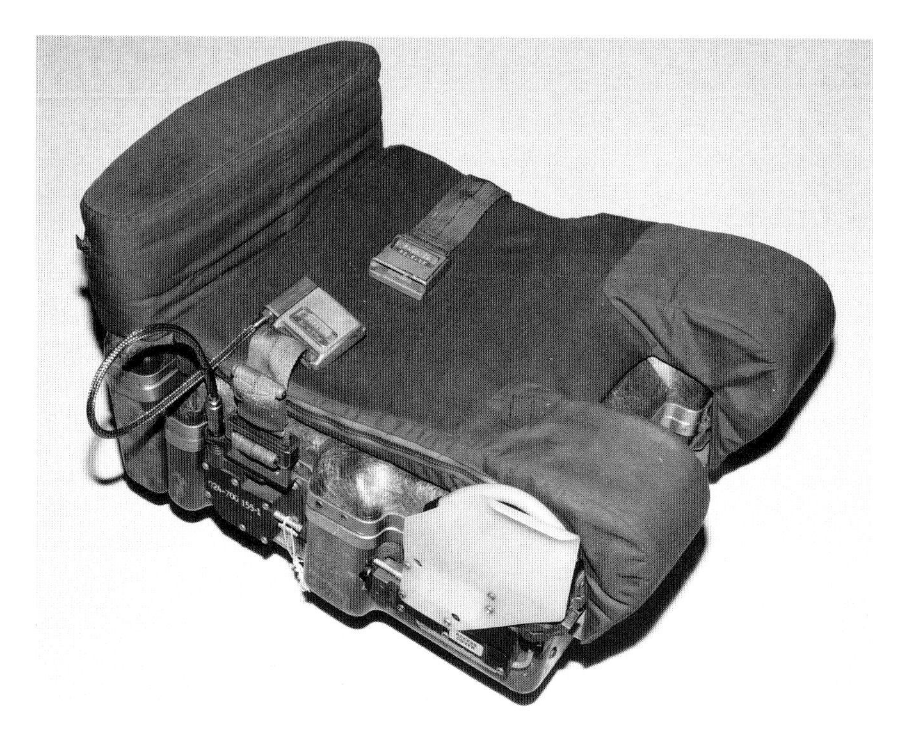

U.S.A.F. F105 Thunderchief seat survival kit

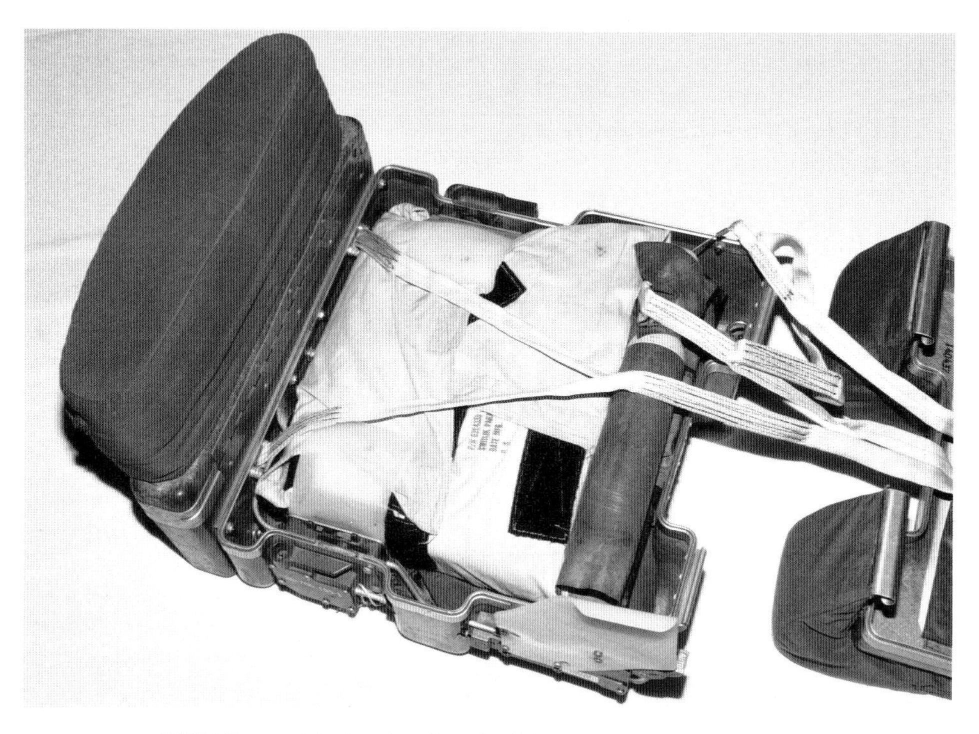

F-105 kit opened to show location of raft (wrapped in protective 'diaper')

EJECTION SEAT SURVIVAL KITS

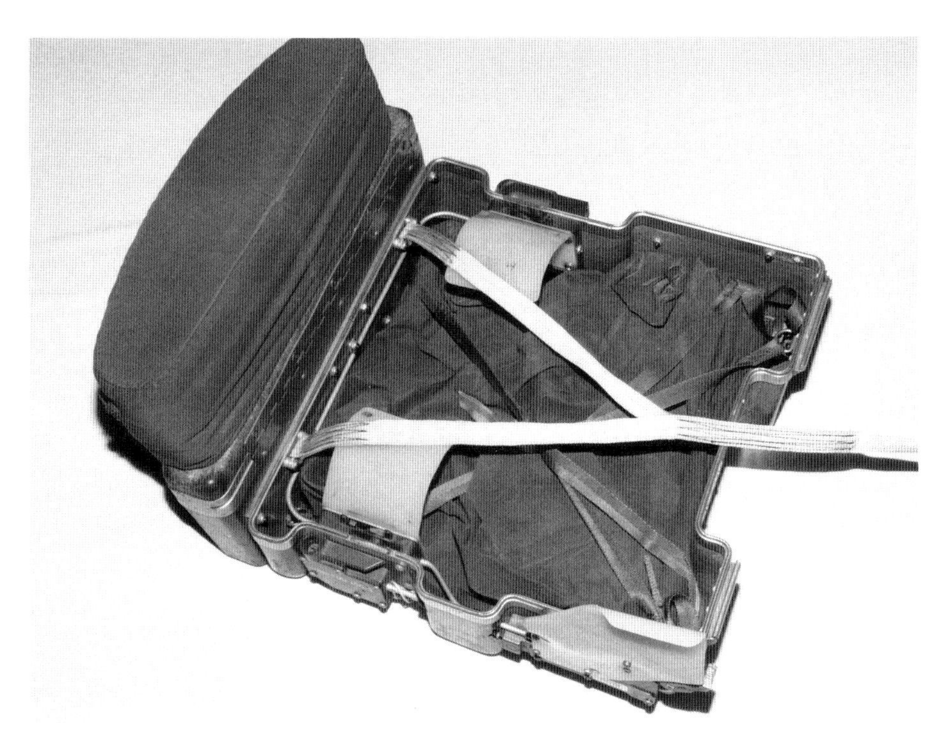

F-105 seat kit showing the inner waterproof container (Raft is on top.)

KIT TYPE	KIT NO	KIT LOCATION	ORGANIZATION DESIGNATION AND LOCATION			
KOCH-61	142241	212 19	149 TFS			
RAFT SN AND MFG DATE		[redacted]	[redacted]	FLARE LOT NO	[redacted]	RADIO/BEACON SN
6737 12/67		LRV 3/P	FIRST AID Dec 81	9-KC-0874	PENGUN-1 COL-50-1	33892
KIT STOCK NO	TECHNICAL ORDER		[redacted]		[redacted]	WEAPON SN
1660 00 169 1733 TP	15X11-19-2		MAY 79		MAY 79	

INSTALLED COMPONENT NOMENCLATURE	AUTH	O/H
RAFT,	1	1
REPAIR PLUGS	2	2
SIGNAL MIRROR MK-3	1	1
SIGNAL FLARE MK-13	2	2
CONTAINER W/MATCHES	1	1
FIRST AID KIT	1	1
WHISTLE	1	1
RADIO 9-78	1	1
SPARE BATTERY FEB 79	1	1
COMPASS LENSATIC	1	1
BLANKET	1	0
BAG WATER STORAGE	1	0
WIRE BRASS	20ft	0
SOCKS	1pr	0
KNIFE SURVIVAL 5"	1	0
GILL NET	1	0
MITTENS W/INSERT	1pr	0
PEN GUN FLARE	1	1
SEA MARKER DYE	1	0
SURVIVAL MANUAL	1	0

INSP DUE	DATE INSP	INSP LAST NAME	TO NUMBER AND DATE
30AUG79	30AUG79	[signature]	15X11-19-507 APR 72
			15X11-19-513 MAY 78
28Dec79	27Dec79	[signature]	15X11-19-511 JUN 79
25APR80	25APR80	[signature]	14S-1-102 JUN 72
23AUG80	8AUG80	[signature]	14S1-3-51 FEB 68
6Dec80	6 Dec 80	[signature]	[redacted]
5APR81	3APR80	[signature]	[redacted]
7AUG81			00-35A-39 MAR 78
			31R2-4-488-1 1Jun73
			11A10-26-7 SEP 75
			11P9-21-7 1Dec75

AFTO FORM JUL 70 338 PREVIOUS EDITION WILL BE USED. SURVIVAL KIT RECORD AFLC–WPAFB–JUL 70–45M

Survival kit record packed in above F-105 seat kit

EJECTION SEAT SURVIVAL KITS

The rear compartment of the F-105 survival seat kit could be used to store extra survival items. (A radio beacon was not installed in this kit.)

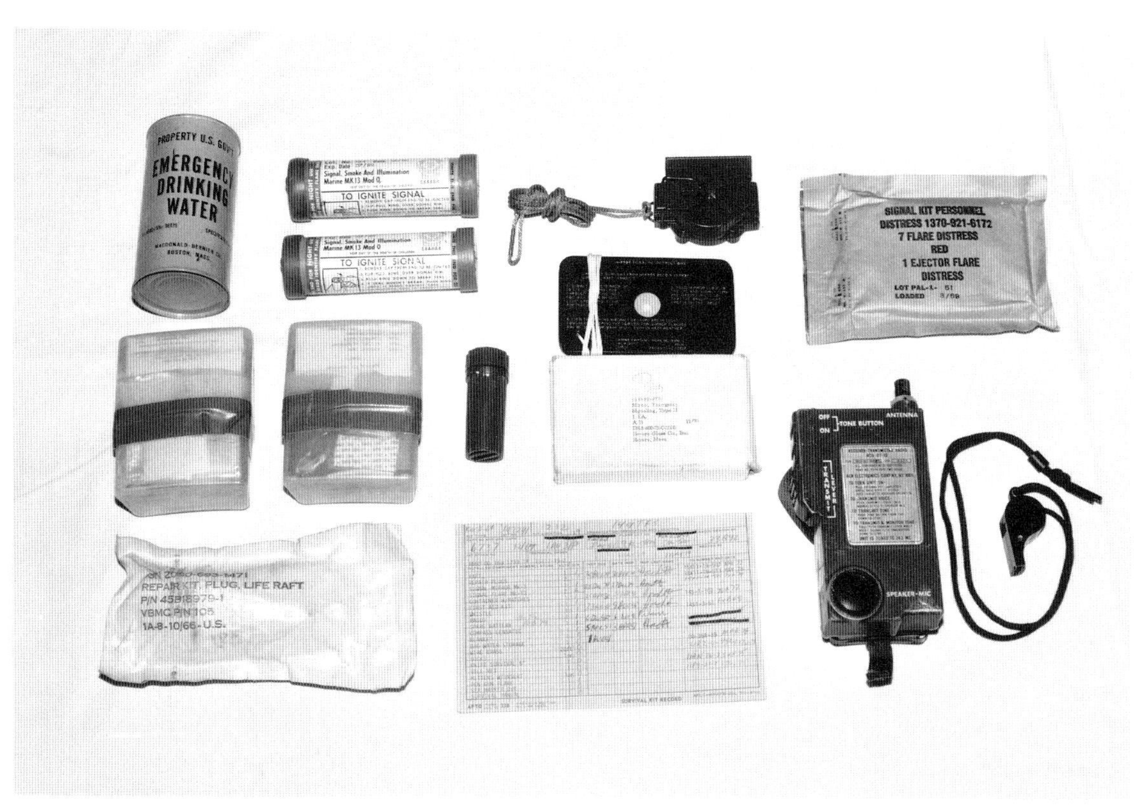

Typical components for an F-105 seat kit (left to right): can of water, 2-part Individual Survival Kit (first aid kit), raft repair kit, Mk-13 Mod 0 smoke/flares, match case, component inventory list, Mk-3 3" x 5" mirror and box, RT-10 radio, whistle, pen flares and launcher in bag

EJECTION SEAT SURVIVAL KITS

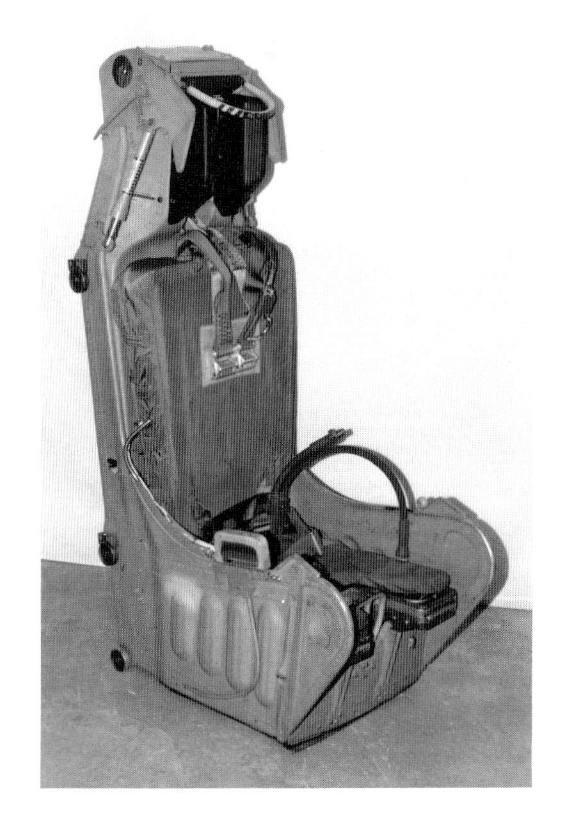

Left: U.S.N. A7E Corsair Escapac IG2 ejection seat

Right: U.S.N. A7E RSSK-8A-1 (Rigid Seat Survival Kit)

LABEL

SURVIVAL KIT CONTAINER, WITH OXYGEN
TYPE RSSK-8A-1
IDENT. NO. 53655-800722-02
SPEC. NO. MIL-S-8101A (AS) & MIL.S-81018/2C(AS)
CONTRACT NO. N00383-741-2279
MFR. SERIAL NO. 1327 DATE 3-75
NATIONAL STOCK NO. 2RM 1660-00-124-1558LX

EJECTION SEAT SURVIVAL KITS

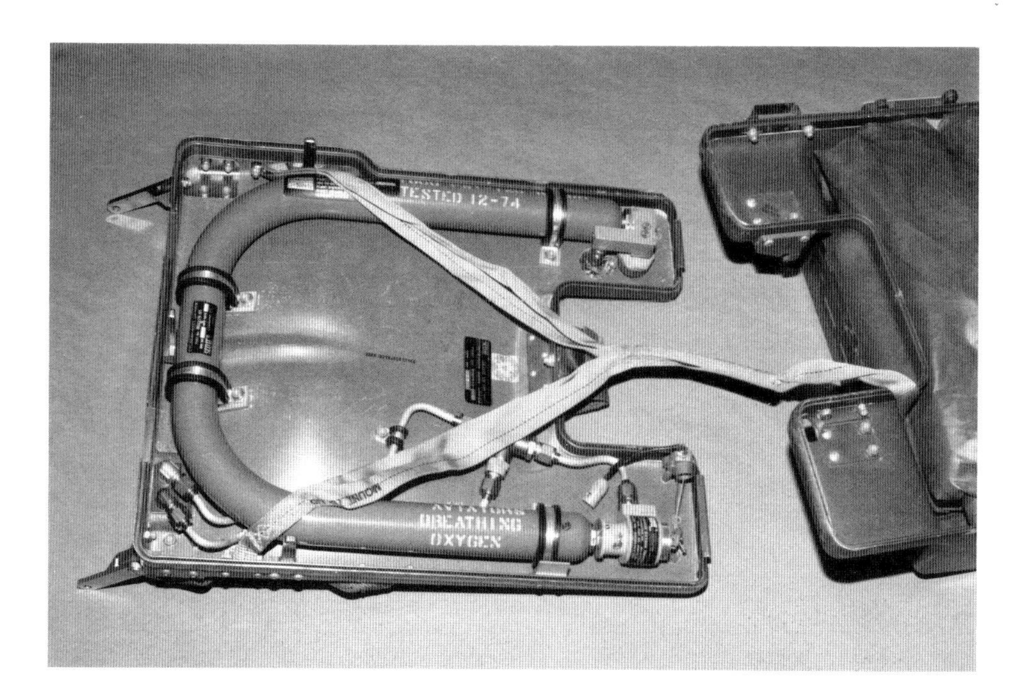

Left: U.S.N. RSSK kit showing oxygen cylinder and drop line attached to lid

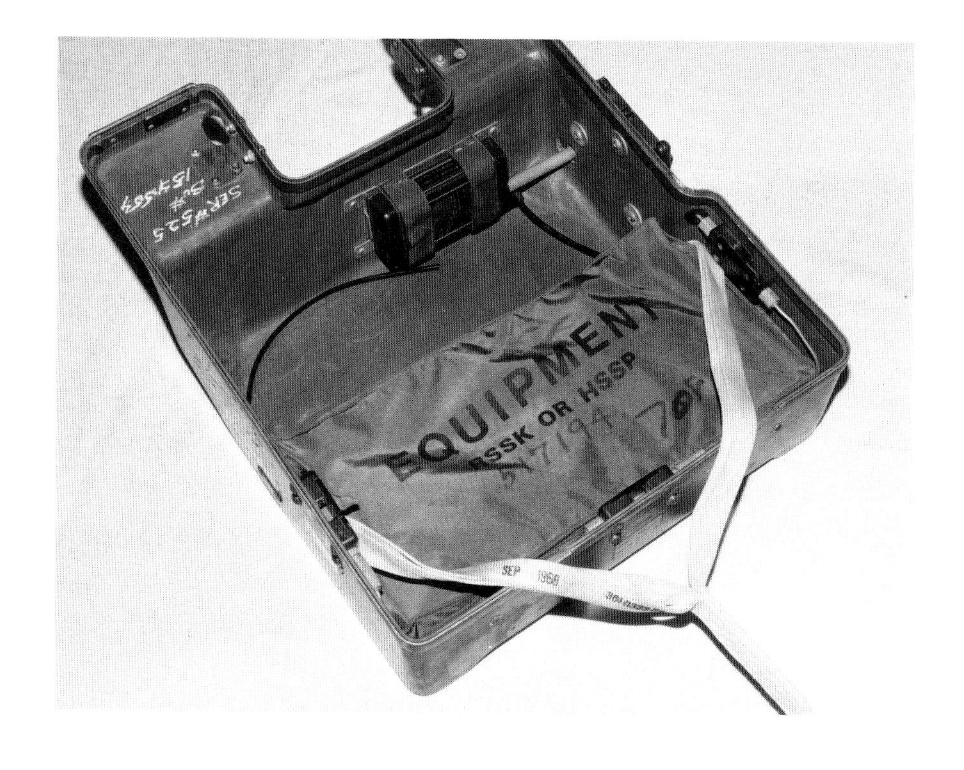

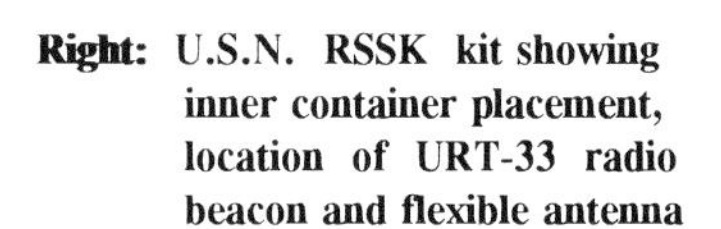

Right: U.S.N. RSSK kit showing inner container placement, location of URT-33 radio beacon and flexible antenna

EJECTION SEAT SURVIVAL KITS

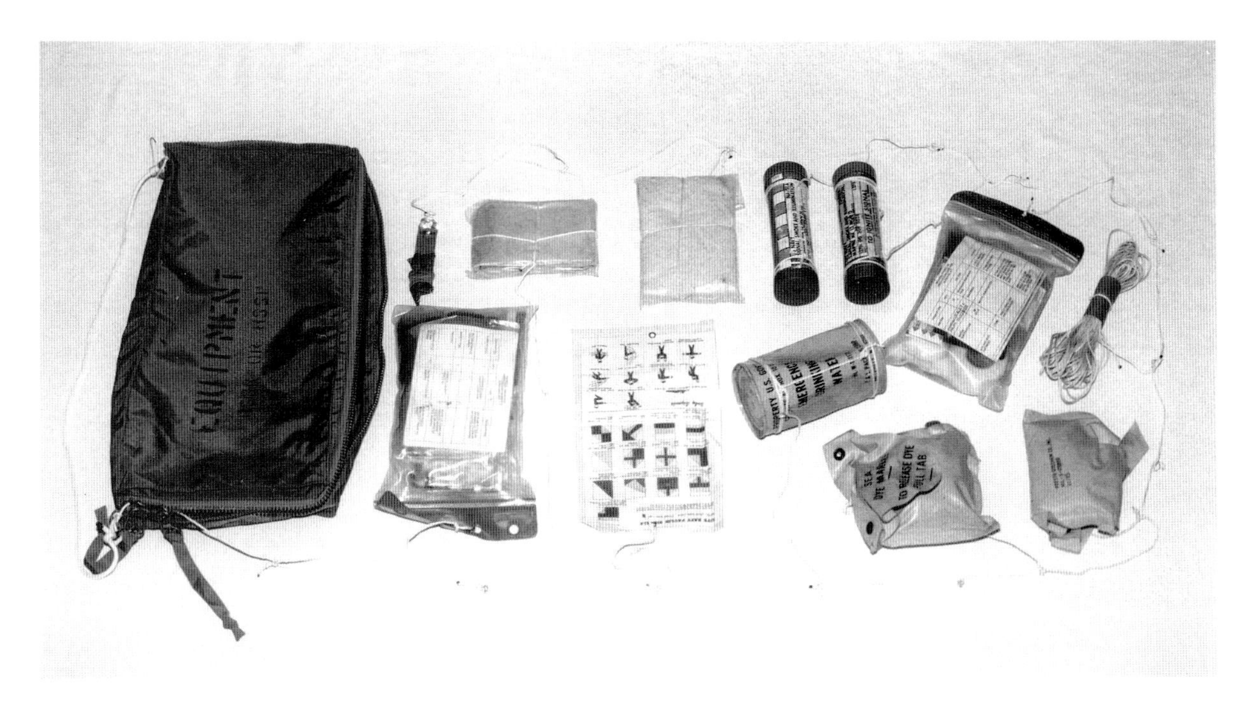

U.S.N. RSSK components (top row, left to right): inner container, can opener (end wrapped to prevent damage to other contents), emergency blanket, sponge, Mk-13 Mod 0 smoke/flares (2), SRU-31/P Medical kit, 50 feet of nylon line (bottom row, left to right): SRU-31/P General kit, code card, can of water, sea dye marker (2) Note: SRU-31/P items are carried in clear ziplock bags instead of the standard green vinyl containers. All items are tied together to prevent loss at sea.

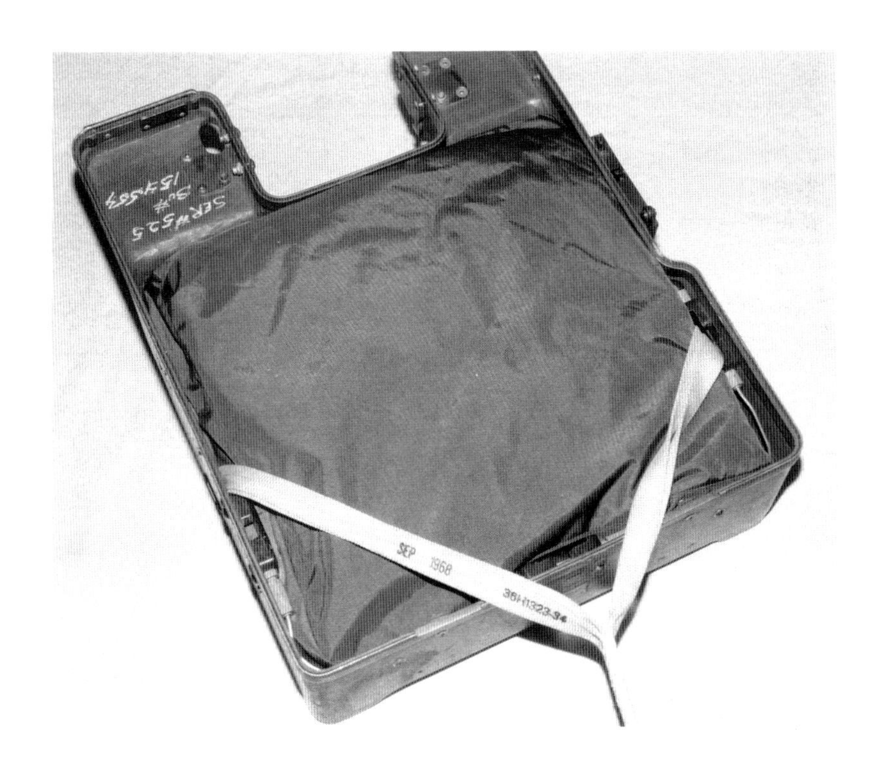

RAFT LABEL (U.S.N. RSSK)

PART NO. 30003-67A318H2-1
LIFERAFT, INFLATABLE, ONE MAN
TYPE LR-1
MIL-L-81542A(AS)
RUBBER CRAFTERS OF W.VA.,INC.
CONT. NO. N00383-82-C-4639
MFG. 4/83
SER. NO. 656

Left: **U.S.N. RSSK kit showing raft protective cover**

EJECTION SEAT SURVIVAL KITS

Left: U.S.A.F. F102 Delta Dagger ejection seat survival kit showing the emergency oxygen hose, communication hookup and 'green apple' oxygen activation ball (An emergency beacon was not installed in this kit.)

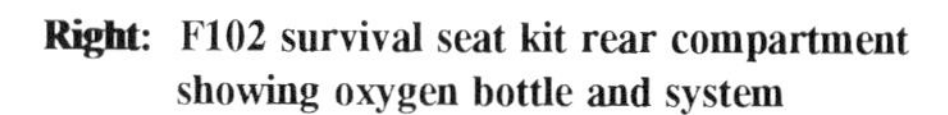

Right: F102 survival seat kit rear compartment showing oxygen bottle and system

LABEL

AUTOMATIC SURVIVAL KIT
CONTAINER
PART NO. 782000-
STOCK NO.
SERIAL NO.
CONTRACT NO. F34601-71-C-3478

EJECTION SEAT SURVIVAL KITS

LABEL

SEAT KIT ASSEMBLY
TYPE RSSK-7
MANUFACTURED BY
CARLETON CONTROLS CORP.
EAST AURORA, N.Y. CODE IDENT. 04577
PART NO. 7192 001-9 SER.NO. 1419
SPEC. NO. GAC128SCES100-9 DATE MFR. 11-74
FED. STOCK NO.
CONTRACT NO. N00018-69-C-0422
TYPE SEAT MK-GRU7 AIRCRAFT F14A
U.S.

U.S.N. F-14A Tomcat ejection seat survival kit Note the early type of kit release handles. (A URT-33 beacon would be used with this kit.)

LABEL

CONTAINER ASSY - SURVIVAL EQUIPMENT,
EJECTION SEAT
PART NO. 88277 - J114510 - 511 CHG 1Y
MFR 88277
CONTRACT NO. F33657 - 82 - C - 2118
SERIAL NO. 3741
DATE PLACED IN SERVICE AUG 1984

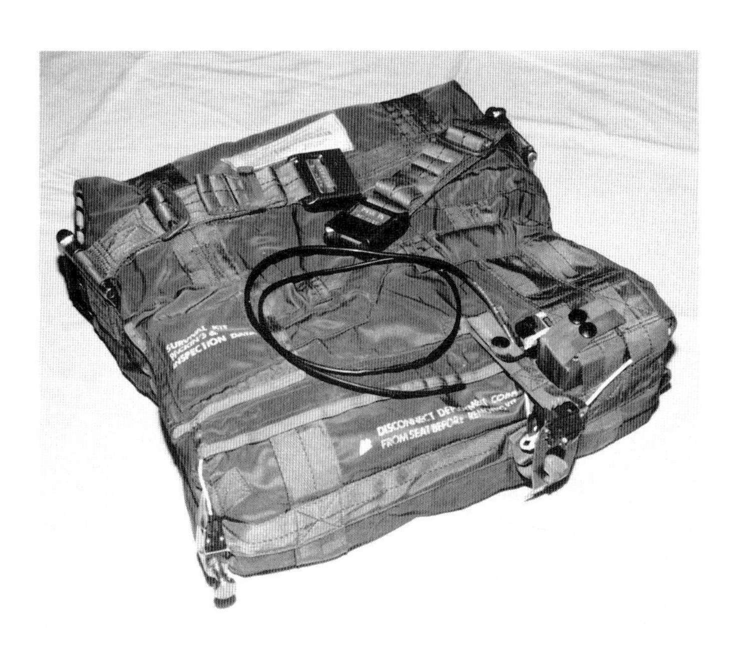

U.S.A.F. F-16 Fighting Falcon ACES II ejection seat survival kit with URT-33C radio beacon installed

MD-1 SEAT KIT

MD-1 seat kit

LABEL

SURVIVAL KIT CONTAINER
CONTOURED SEAT
TYPE MD-1
PART NO. 48F3865
ORDER NO. AF09(603)61736
SWITLIK PARACHUTE CO., INC.
PROPERTY U.S. AIR FORCE

The MD-1 container is a seat kit used in conjunction with back and chest style parachutes in both non-ejection and ejection seat aircraft such as the F-80, F-84, F-86, F-100, F-101, B-47 and early B-52. The container consisted of an MA-1 seat cushion, magnesium seat pan (earlier models had fiberglass pans) attached to an expandable fabric body, waterproof inner container, 25-foot drop lanyard and attachment straps. The container is closed by means of a slide fastener that is locked by a quick removable barrel keeper, and is attached to the pull tab. The fabric section is snugged up tightly with a drawstring around the bottom edge. The outer container provides storage space for the inner container, with necessary survival components, and the one-man life raft. The inner container and the raft are attached to the outer container by a 25-foot drop lanyard. The outer container attaches to the parachute harness accessory rings with quick release fasteners.

The MD-1 kit is opened by pulling the zipper pull tab at a 90 degree right angle, directly away from the container. This pulls the barrel keeper off the slide fastener and activates the liferaft C02 inflation system which inflates the raft and in turn, completes opening the container.

MD-1 SEAT KIT

The MD-1 container is similar to the 'Container Assembly, Emergency Sustenance Kit, Parachute Bailout Seat Style' described in versions of AFM 64-4 and the Class 20-B manuals of the early 50's. Four different kits were listed depending upon the type of climate to be encountered but used the MD-1 type container. The A-1 kit was for very cold climates, B-1 kit was for cold climates, C-1 kit was for use in temperate zones and the F-1 kit was used in hot climates. Components were selected to best suit the climate.

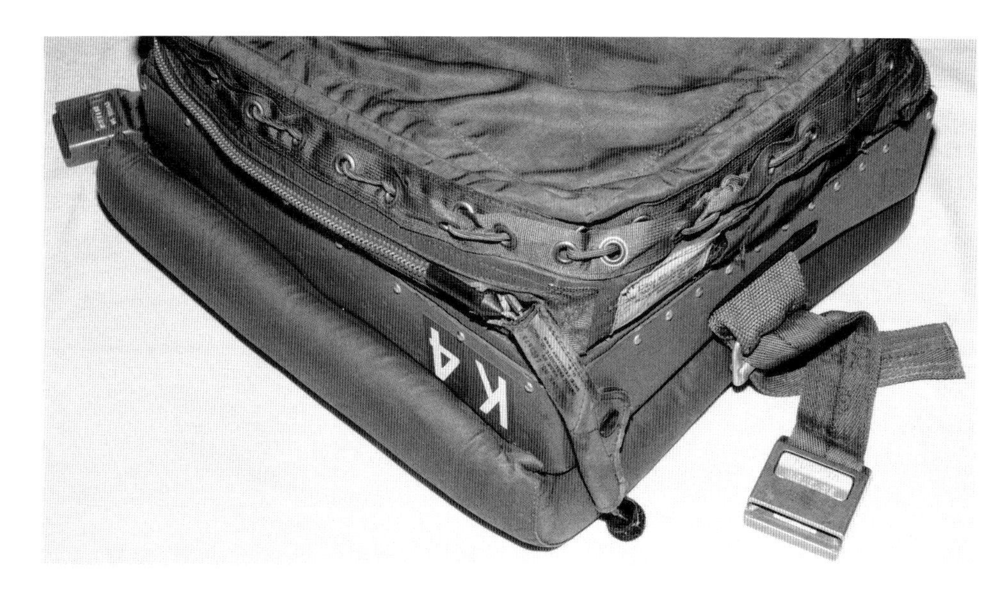

Right: MD-1 seat kit, bottom view, showing equipment release pull knob

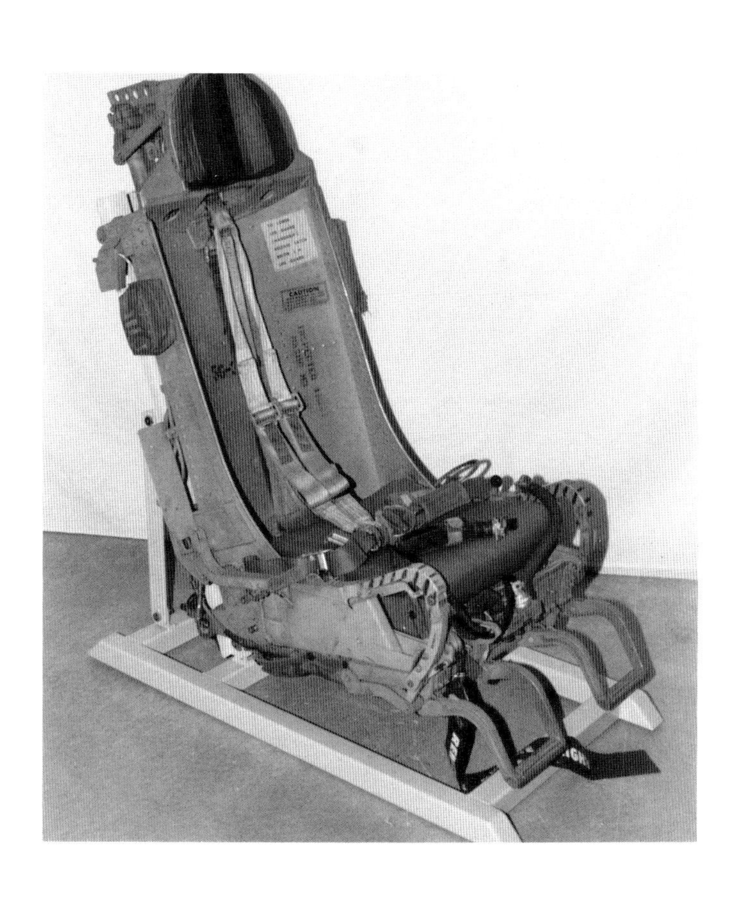

Left: The U.S.A.F. F100F Super Sabre ejection seat uses a MD-1 seat survival kit. The radio beacon and oxygen cylinder would be carried in the back style parachute.

MB-2 SEAT KIT

MB-2 seat kit

LABEL

SURVIVAL KIT CONTAINER
DWG.-55K3976
S/N 4220-565-3277
LITE INDUSTRIES, INC.
CONT. NO. N383(MIS)61564A
U.S.

LABEL ON SEPARATOR

P/N 55C3079
SWITLIK PARACHUTE CO., INC.
DATE MFG. SEPT. 1961

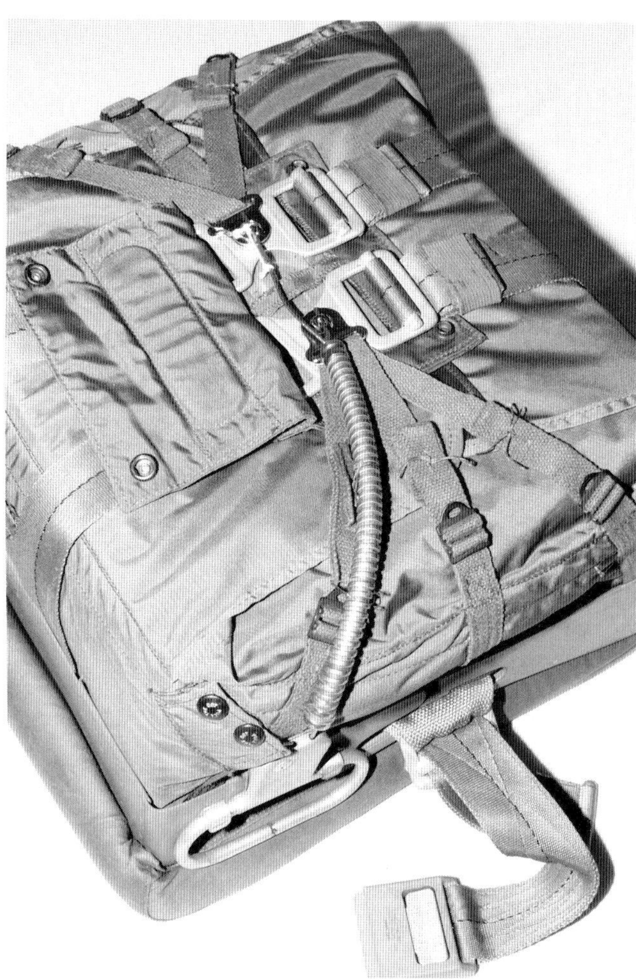

Right: MB-2 seat kit, bottom, showing release arrangement

The type MB-2 seat kit is similar to the MD-1 seat kit but incorporates a separator, which divides the inner container and raft from the closing flaps, and a ripcord assembly which is used to open the container during emergency use. The MB-2 also uses the type MA-1 cushion. The container is closed by inserting pins attached to the ripcord assembly into cones on the fabric closing flaps. The MB-2 seat kit also uses a 25-foot drop line to attach the liferaft and inner container to the outer container. The kit is opened by pulling the 'D' ring mounted on the right side of the seat pan. This opens the fabric closing flaps and activiates the raft CO2 cylinder. When the container is opened, the raft and inner container drop out and the raft inflates. The raft and container hang below the crewmember on the drop lanyard.

The MB-2 seat kit could not be packed to obtain the correct angle for proper ejection seating and was not recommended for use in certain ejection seat aircraft such as the T-38, B-58, F-100, F-101, F-102, F-104, F-105, F-106 and later B-52's. However, it could be used in certain fighter, bomber and transport aircraft and was listed as a substitute for the MD-1 and ML-4 seat kits.

U.S.A.F. SEAT KIT COMPONENTS

The mandatory items listed herein are in addition to those items assumed to be carried outside the kit in or on the clothing or parachute, such as the URT-21/27/33 beacon, SDU-5/E distress light, A/P25S-1 distress kit, anti-exposure suits (as necessary), SRU-16/P minimum kit and the MC-1 knife. If after installation of mandatory items and those by direction of Commanders additional space is available, components should be chosen from the optional list of items that will be useful in various types of conditions. Desirable and 'nice to have' type components shall not be given consideration since they only tend to cause kit overpacking and subsequent kit failure. When choosing items, use the philosophy of life protection items; signal device; and escape and evasion items.

The mandatory and optional items listed are applicable to all individual hard and soft containers. The sleeping bag requirement is exempt for the ML-4 and other seat kits where space is not available.

MANDATORY ITEMS

ITEM	STOCK NUMBER
-Life raft, Type LRU-3/P (1)	4220-726-0424LS
-Life raft, Type LRU-6/P (1)	4220-869-2738LS
-Life raft, Type LRU-4/P (1)	4220-132-4230LS
-Raft repair kit (1)	4220-763-3766LS
* -Sleeping bag (1)	
@ -Mk-13, Mod. 0, Marine smoke and illumination signal (2)	1370-309-5027-L275
-Mk-3, Type II mirror (3" x 5") (1)	6350-299-6197
-Mk-3, Type I mirror (2" x 3") (1)	6350-NSL
-Packet sea marker (1)	6850-270-9986
-Matchbox, waterproof/w matches (1)	6865-265-4925
** -First aid kit (1)	
-Whistle, Police, plastic (1)	8465-254-8803
***-ACR/RT-10 radio set (1)	5821-912-4480

OPTIONAL ITEMS

ITEM	STOCK NUMBER
-Radiacmeter IM-179 (1) or	6665-975-5167
Radiacmeter (1)	6665-679-2879
-Survival manual (1)	AF Manual 64-5
-Canned drinking water (1)	8960-243-2103
-Socks, wool, ski (1 pr.)	8440-153-6717
-Goggles, Sun, Type MA-1 (1 pr.)	8465-530-4083
@@ -Food packet, Survival General Purpose (2)	8970-082-5665
-Cartridge, ball, caliber 22, Hornet (50)	1305-529-8878-A087
-Shark deterrent (1)	6850-281-6926
-Gill net, nylon (1)	4240-300-2138
-Rifle, survival, M-4, 22 Hornet (1)	1005-575-0070
-Wire, Comm. brass	9525-596-3498

U.S.A.F. SEAT KIT COMPONENTS

OPTIONAL ITEMS - Continued

ITEM	STOCK NUMBER
-Candle, Long burning (1)	6260-840-5578
-Heat tabs, J. W. Speaker, P/N 1118 w/stove (1 box of 24) w/stove (1 box of 24)	L.P. - J. W. Speaker Corp.
-Saw, hand, finger grip Type MD-2 (1)	5110-570-6896
-Knife, pocket (1)	7340-162-2205
-Compass, lensatic	6605-846-7618
-Bag, storage, drinking water, Size B, (3 pint)	8465-634-4499
-Desalting kit (1)	4220-216-5031
-Magazine, 22 Hornet (1)	1005-726-5895
-Hood, winter (1)	8415-543-7130
-Signal kit, personal distress Type A/P25S-1	1370-921-6172-LY35
-Drawers, waffle weave series (1 pr.)	8415-782-3226
-Undershirt, waffle weave series (1)	8415-270-2012
-Gloves, inserts (1 pr.)	8415-269-0500
-Mittens, heavy, N4B (1 pr.)	8415-268-8312
-Boot, Mukluk 12 inches high adjustable size (1 pr.)	8430-275-5732
-Insect repellent, stick form 6-12 or equal (1)	Commercial
-Hat and mosquito net (1)	8415-269-0492
-Ointment, sun (1)	8510-162-5658
-Rifle, shotgun, M-6, 22 Hornet, 410 gage (1)	1005-575-0073
-Shell, 410 gage, 3 in. case (1 box)	1305-028-6644-A055
-Starter, fire, Type M2 (2)	1370-219-8566-L621
-Soap, toilet, bar (1)	Commerical
-File, flat 6 in., Type B (1)	5110-234-6532
-Survival tool kit, Type SRU-18/P (1)	4240-065-6713
-Plastic, sheet rough translucent (6 ft.)	Commerical
-Bag, motion sickness (2)	8105-122-3385
-Aluminum blanket (1)	L.P. National Research Corp.
-Knife, hunting with sheath (1)	7340-292-9138

* There are three different types of sleeping bags for use in various types of aircraft and survival kits. These are:

FSN 8465-753-6325
FSN 8465-753-6143
FSN 8465-753-3226

Sleeping bag to be utilized shall be determined by type of aircraft/containers, or by major air commands for particular mission.

** There are a number of first aid kits suitable for use with individual kit. Either of the following is recommended:

FSN 6545-823-8165 First Aid Kit , Individual
FSN 6545-611-0978 Two Part First Aid Kit
FSN 6545-782-6412 Tropical Personal Aid Kit 2D/1 TAC

U.S.A.F. SEAT KIT COMPONENTS

*** Radios will be installed only in those kits used on Tactical Aircraft of commands having a mission requiring downed personnel voice authentication. Commands having a requirement for only one radio per crew position have the option of installing it in either the survival kit or the survival vest.

@ Pengun type flares may be substituted for these flares with using Command approval.

@@ NATO mandatory

B-58 ESCAPE CAPSULE

B-58 'Hustler' survival kit containers

Unlike the open ejection seat, the unique, even bizarre, B-58 'Hustler' aircraft escape capsule was a completely automatic survival system. The rocket type escape capsule at each crew station permitted safe ejection from the aircraft at high speeds and high altitudes. Each capsule was equipped with three airtight clamshell doors, a window, independent pressurization and oxygen supply systems (normal and emergency), seat and back cushions, headrest, restraint equipment, shoulder harness, inertia reel, parachute, stabilization equipment and survival equipment. The survival components were vacuum packed and carried in several specifically molded fiberglass containers that fit into various sections of each capsule. The pilot capsule had six containers and crew member capsules numbered eight. The survival items that were provided for the flight crew were some of the most extensive and specialized individual components ever issued.

On land the escape capsule could be used as a shelter to protect the survivor against all elements and natural enemies. On the desert, it provides protection against sand storms while in the north it provides protection against blizzards.

Special artic clothing provided in the survival kits enhanced survival chances in temperatures as low as 60 degrees below zero. The survial kit included a three day water and a fourteen day food supply.

B-58 ESCAPE CAPSULE

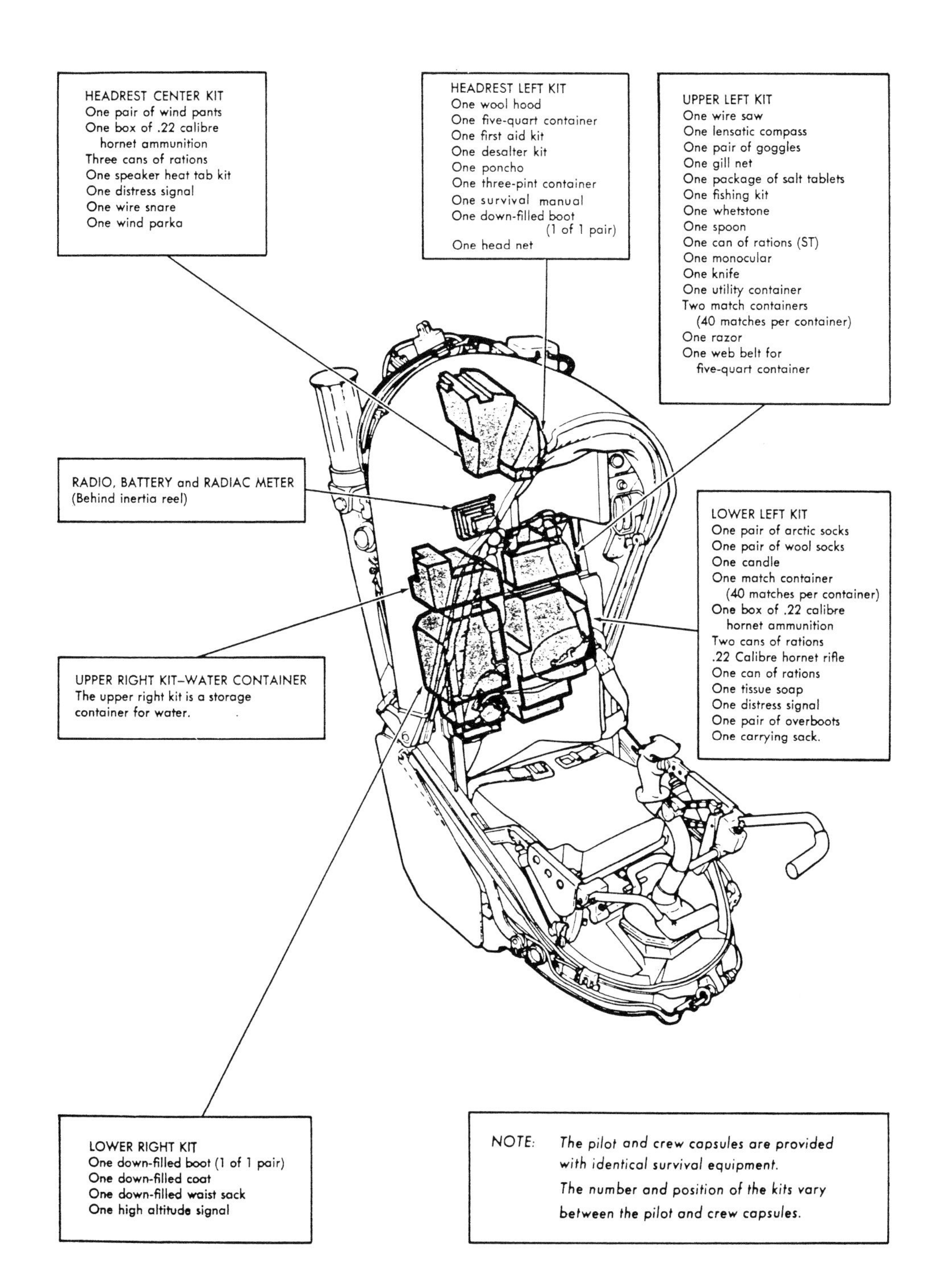

HEADREST CENTER KIT
One pair of wind pants
One box of .22 calibre hornet ammunition
Three cans of rations
One speaker heat tab kit
One distress signal
One wire snare
One wind parka
HEADREST LEFT KIT
One wool hood
One five-quart container
One first aid kit
One desalter kit
One poncho
One three-pint container
One survival manual
One down-filled boot (1 of 1 pair)
One head net
UPPER LEFT KIT
One wire saw
One lensatic compass
One pair of goggles
One gill net
One package of salt tablets
One fishing kit
One whetstone
One spoon
One can of rations (ST)
One monocular
One knife
One utility container
Two match containers (40 matches per container)
One razor
One web belt for five-quart container
RADIO, BATTERY and RADIAC METER (Behind inertia reel)
LOWER LEFT KIT
One pair of arctic socks
One pair of wool socks
One candle
One match container (40 matches per container)
One box of .22 calibre hornet ammunition
Two cans of rations
.22 Calibre hornet rifle
One can of rations
One tissue soap
One distress signal
One pair of overboots
One carrying sack.
UPPER RIGHT KIT–WATER CONTAINER
The upper right kit is a storage container for water.
LOWER RIGHT KIT
One down-filled boot (1 of 1 pair)
One down-filled coat
One down-filled waist sack
One high altitude signal
NOTE: The pilot and crew capsules are provided with identical survival equipment.
The number and position of the kits vary between the pilot and crew capsules.

Pilot's Escape Capsule

B-58 ESCAPE CAPSULE

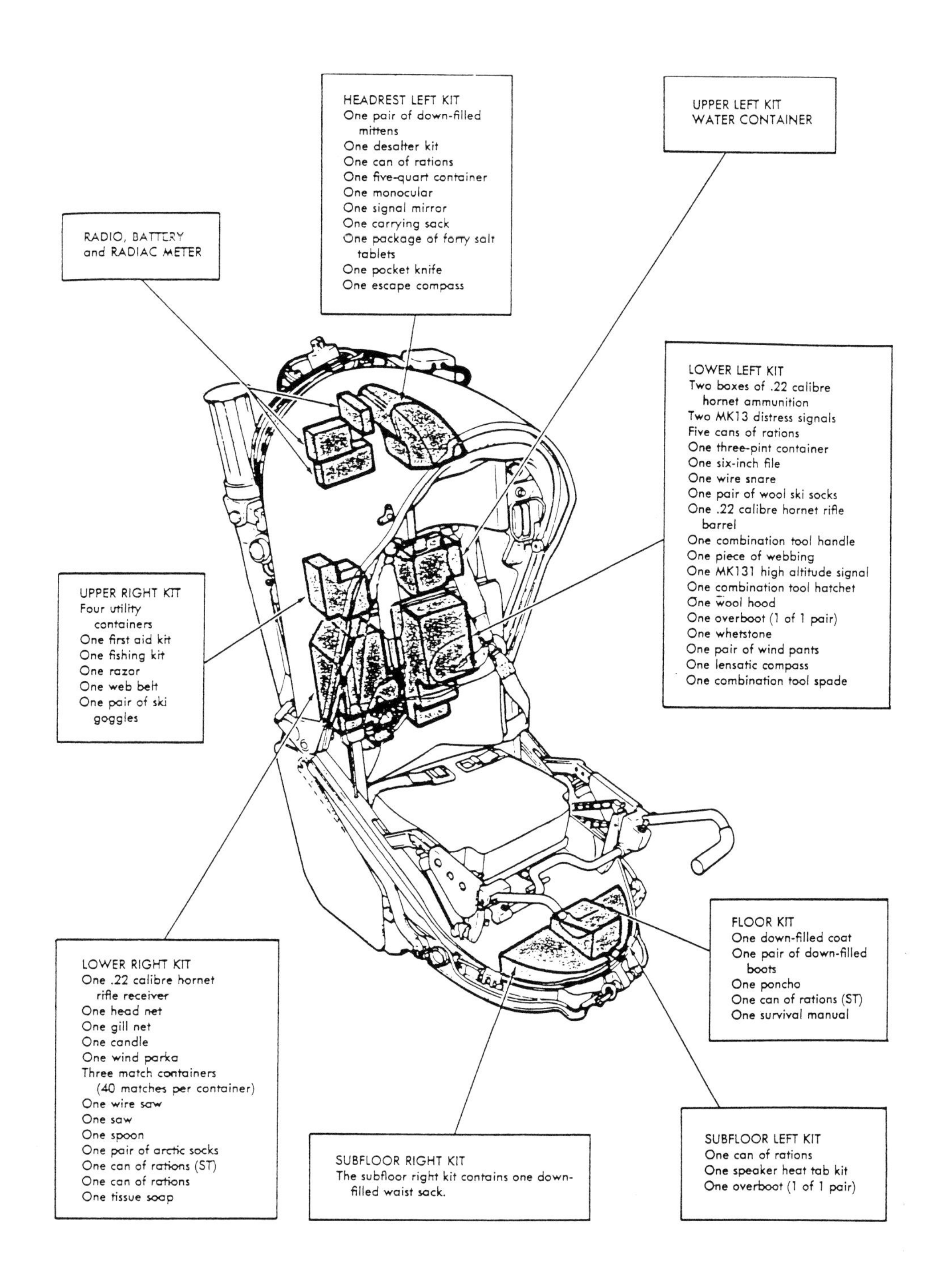

HEADREST LEFT KIT
One pair of down-filled mittens
One desalter kit
One can of rations
One five-quart container
One monocular
One signal mirror
One carrying sack
One package of forty salt tablets
One pocket knife
One escape compass
UPPER LEFT KIT
WATER CONTAINER
RADIO, BATTERY and RADIAC METER
LOWER LEFT KIT
Two boxes of .22 calibre hornet ammunition
Two MK13 distress signals
Five cans of rations
One three-pint container
One six-inch file
One wire snare
One pair of wool ski socks
One .22 calibre hornet rifle barrel
One combination tool handle
One piece of webbing
One MK131 high altitude signal
One combination tool hatchet
One wool hood
One overboot (1 of 1 pair)
One whetstone
One pair of wind pants
One lensatic compass
One combination tool spade
UPPER RIGHT KIT
Four utility containers
One first aid kit
One fishing kit
One razor
One web belt
One pair of ski goggles
FLOOR KIT
One down-filled coat
One pair of down-filled boots
One poncho
One can of rations (ST)
One survival manual
LOWER RIGHT KIT
One .22 calibre hornet rifle receiver
One head net
One gill net
One candle
One wind parka
Three match containers (40 matches per container)
One wire saw
One saw
One spoon
One pair of arctic socks
One can of rations (ST)
One can of rations
One tissue soap
SUBFLOOR RIGHT KIT
The subfloor right kit contains one down-filled waist sack.
SUBFLOOR LEFT KIT
One can of rations
One speaker heat tab kit
One overboot (1 of 1 pair)

Crew's Escape Capsule

B-58 ESCAPE CAPSULE

B-58 'Hustler' right hand external crew container (center): showing all items in sealed vacuum packed container (loose, left to right): 6" file,button escape compass and container, roll of snare wire, spade, foot sack

OV-1 EJECTION SEAT KITS

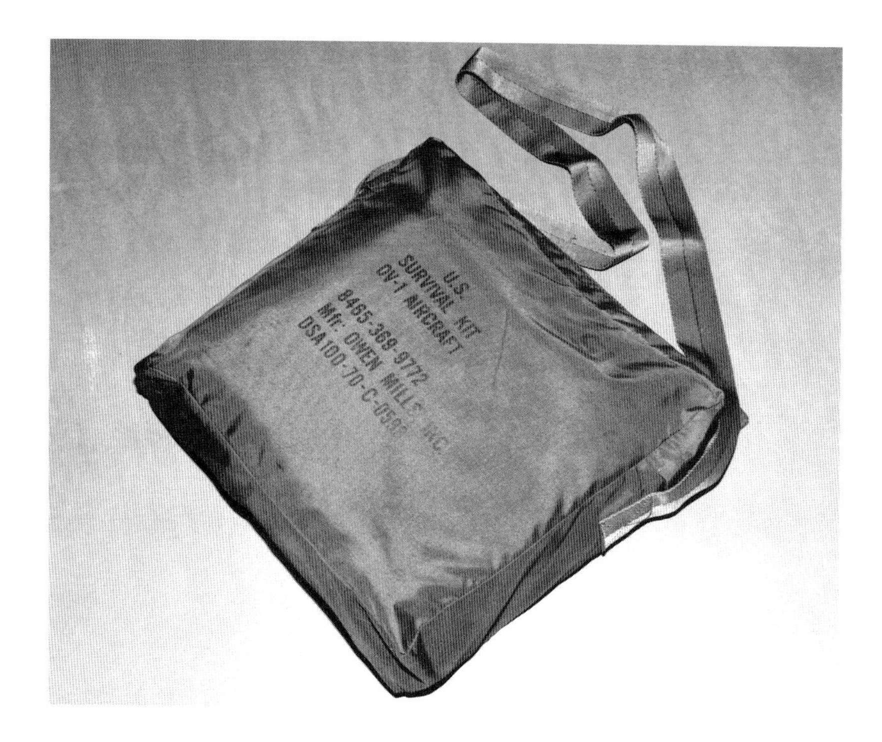

OV-1 Hot Climate 'soft' type survival kit

LABEL

U.S.
SURVIVAL KIT
OV-1 AIRCRAFT

8465-369-9772
Mfr: OWEN MILLS, INC.
DSA 100-70-C-0598

The OV-1 Army Mohawk aircraft use a 'soft' type individual survival kit in earlier ejection seats and a Rigid Seat Survival Kit (RSSK) in the later Martin-Baker ejection seats. The 'soft' kits have an outer case made of either canvas or nylon, which fits into the seat pan of the ejection seat and is designed for either hot, cold or overwater environments. The hot and cold climate kits are similar in appearance. The kits measured 14" x 13" x 5" and weighes about seventeen pounds. A web carrying strap attaches to both sides of the case and is stored in the case until required. The hot and cold climate containers have a zipper closure on one side only. The overwater kit has two compartments, one containing the PK-2 one-man raft and CO2 cylinder, and the other containing an inner container with the survival gear. The overwater kit is equipped with a small web carrying handle on one end and two adjustable yellow web shoulder straps which attaches to one side of the case. This kit measures 15" x 13" x 5" and weighes 28 pounds. It is similar in appearance and design to the U. S. Navy PK-2 Pararaft Kit.

The hot and cold climate kits also contained a 12 1/2" x 13 3/4" x 2" aluminum frying pan and the cold kit uses a modified food packet instead of the standard G.P. ration.

OV-1 EJECTION SEAT KITS

The OV-1 Rigid Seat Survival Kits (not shown) are similar except that the container is made of rigid fiberglass and includes a retention assembly for the seat occupant. The survival kit assembly fits in the seat pan and is secured by two lower attachment lugs, that are released only when the time release mechanism or the manual override handle is actuated. An occupant retention assembly is attached to both sides of the survival container and clips to a floating lap belt which is attached to the personal harness. To gain access to an equipment bag containing the survival gear, two special grip handles must be squeezed and pulled to release the assembly which locks the container bottom to the lid assembly.

The cold and hot climate RSSK's weigh thirty-three pounds and measure 15" x 14 1/2" x 9". The overwater kit has the same dimensions but weighs forty-four pounds.

Because the OV-1 survival seat kits are used in conjunction with the OV-1 survival vest, first aid kits are not carried in the seat containers.

Components for these kits are similar to those used in the Army individual survival kits and are listed in the assembly sheets in the appendix.

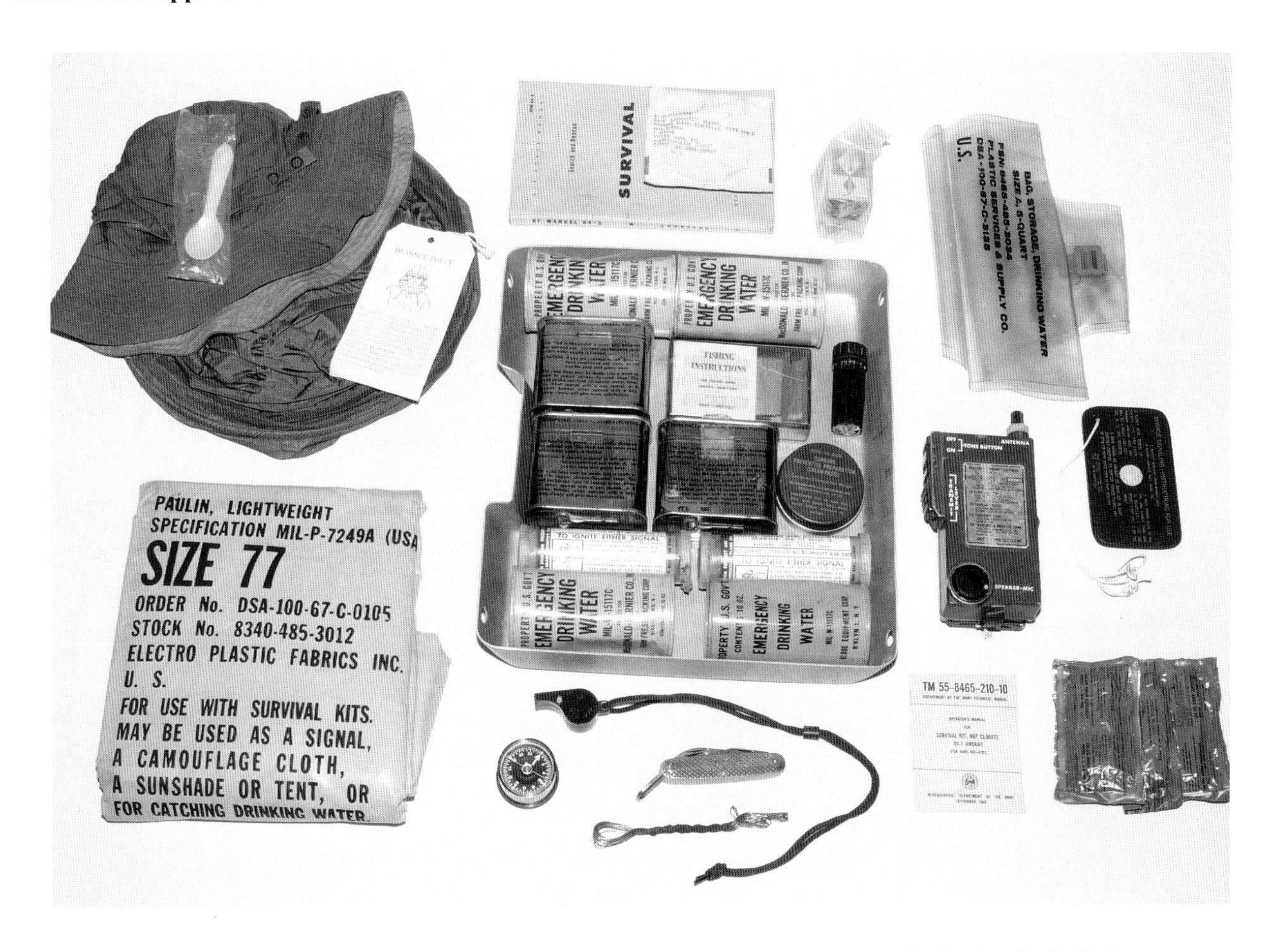

OV-1 Hot Climate Kit (left to right): reversible colored hat, spoon, mosquito headnet (under hat), paulin, whistle, MC-1 compass, snare wire, pocket knife, frying pan, cans of water (4), rations (3), Mk-13 Mod 0 smoke/flares (2), fishing kit, match case, sunburn ointment, survival manual, finger saw in bag, matches in bag, 5 quart water bladder, RT-10 radio, operators manual, trioxane fuel packets, Mk-3 3" x 5" mirror

OV-1 EJECTION SEAT KITS

OV-1 Overwater 'soft' type survival kit

LABEL

U.S.
SURVIVAL KIT OVER-WATER
8465-782-3032
Mfr: OWEN MILLS, INC.
DSA 100-69-C-1864

LABEL

LIFERAFT
ONE MAN INFLATABLE
TYPE PK-2
USN
MIL-L-86648(WP)
RUBBERCRAFT, DIV. KMS IND.
DSA 700-69-C-9377
DATE JUNE 1969
SERIAL NO. 253801

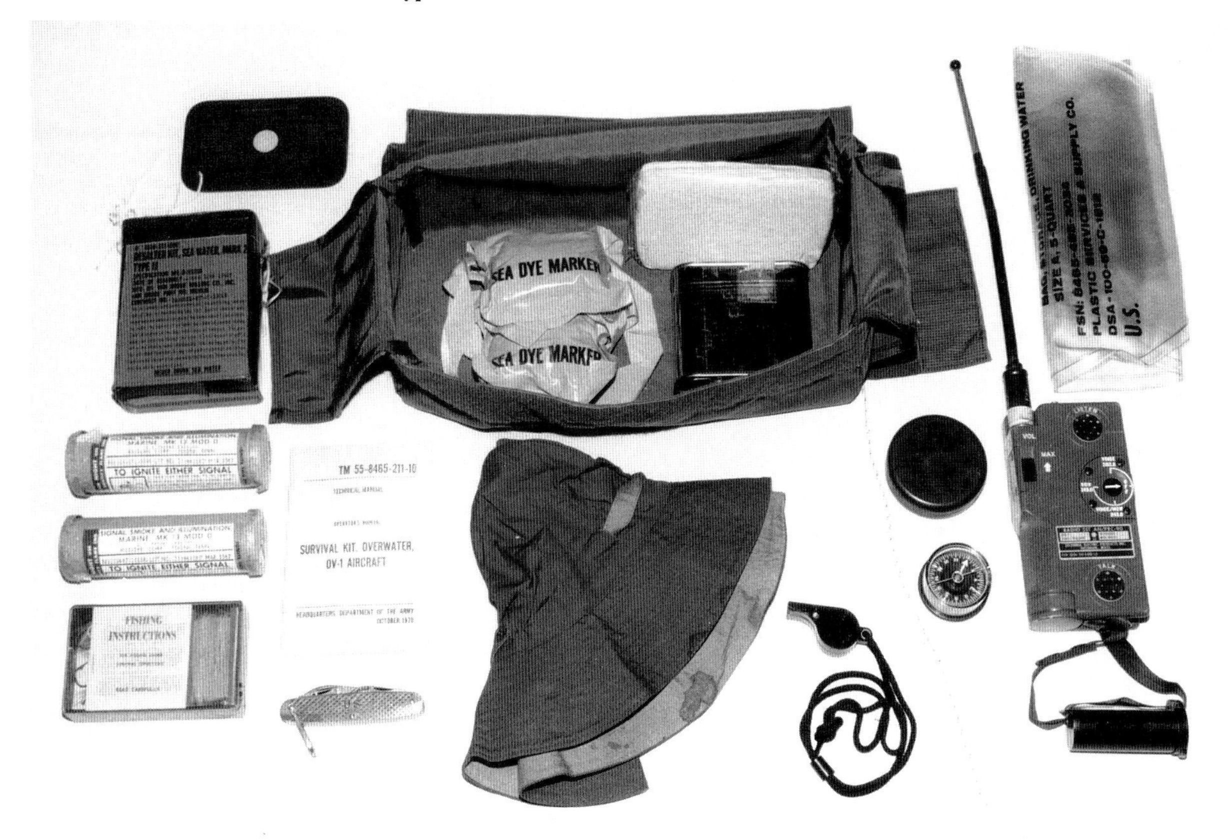

OV-1 Overwater Kit (left to right): fishing kit, Mk-13 Mod 0 smoke/flares (2), desalter kit, Mk-3 3" x 5" mirror, inner nylon container, sea dye marker (2), sponge, ration can, operators manual, pocket knife, reversible hat, whistle, MC-1 compass, sunburn cream, PRC-90 radio, 5 quart water bladder

OV-1 EJECTION SEAT KITS

LABEL

SURVIVAL KIT
OV-1 AIRCRAFT
COLD CLIMATE
FSN 8465-782-3030
AIRFLOTE INC.
DSA 100-67-C-1602

Left: **OV-1 Cold Climate 'soft' type survival kit**

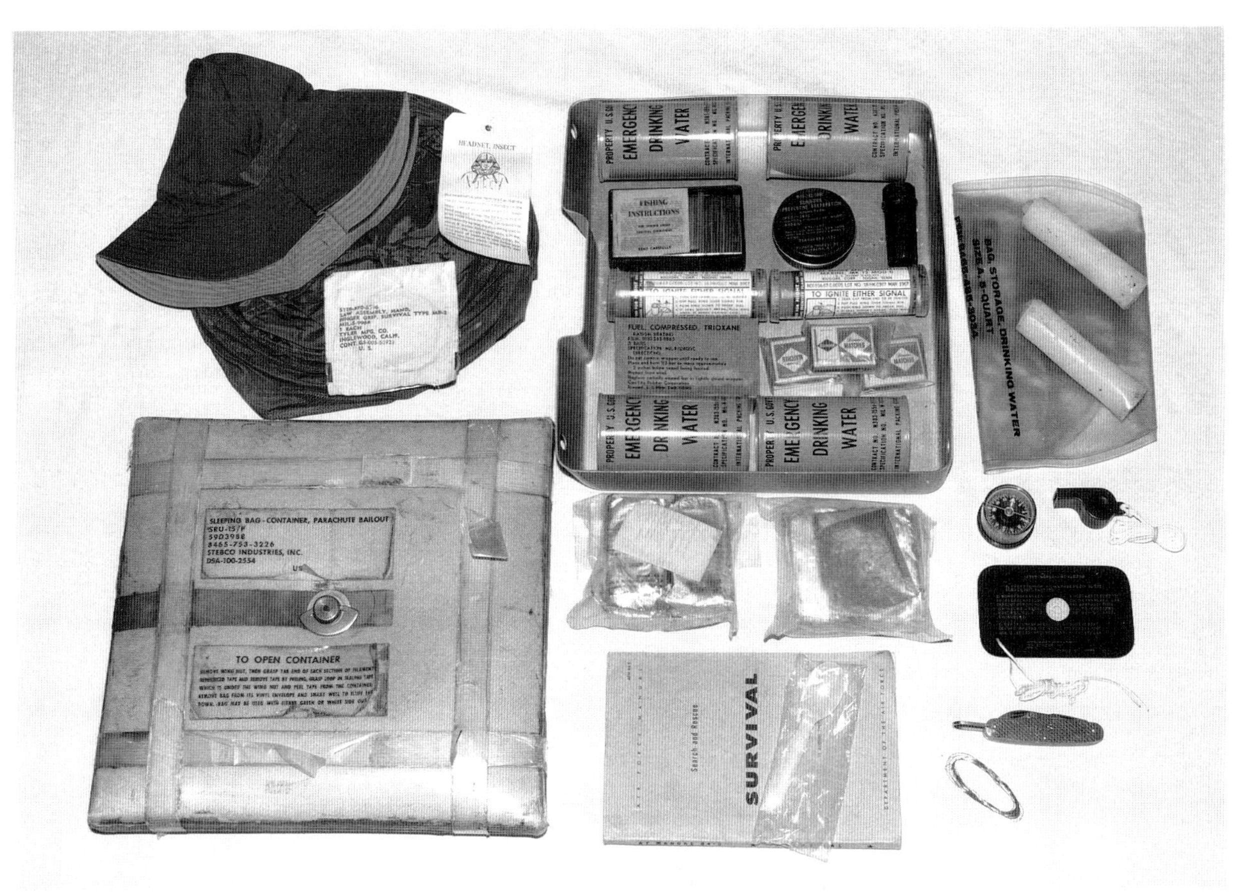

OV-1 Cold Climate Kit components (left to right): reversible hat, mosquito headnet, finger saw in bag, SRU-15/P sleeping bag, survival manual, spoon, rations (modified packets), frying pan, cans of water (4), fishing kit, matches, trioxane fuel, Mk-13 Mod 0 smoke/flares (2), match case, sunburn ointment, 5 quart water bladder, MC-1 compass, candles, whistle, Mk-3 3" x 5" mirror, pocket knife, snare wire

CHAPTER 3

PERSONAL SURVIVAL KITS AND FIRST AID KITS

Personal Survival Kits (PSK's) were in limited use before World War II and became a standard issue item during and after that war. PSK's and individual survival kits differ from other types of survival kits in that they can be a component of the survival vest, seat and back pad kits or can be used totally on their own. They are usually small enough to fit in a pocket or hung from a belt, and except for the first aid kits, contain a variety of medical, sustenance and survival items that can support an individual for a minimum period of time.

PSK's were designed to contain their components in either one or two containers and many individuals created their own personnel kits for their specific needs.

First aid kits (containing only medical items) have been carried on the individual, in survival vests and seat kits since World War II. Unlike the PSK's which contain some medical and survival gear, the first aid kits had only medical items and were located for quick emergency use or to compliment other vest or seat kit survival equipment.

Snake bit and insect sting first aid kits were also carried in vests and kits if the geographical area warranted their use.

An Army Air Force aircrew member uses the small compass from his E-17 'personal aids' kit.

F-1 EMERGENCY SUSTENANCE FLYERS CASE

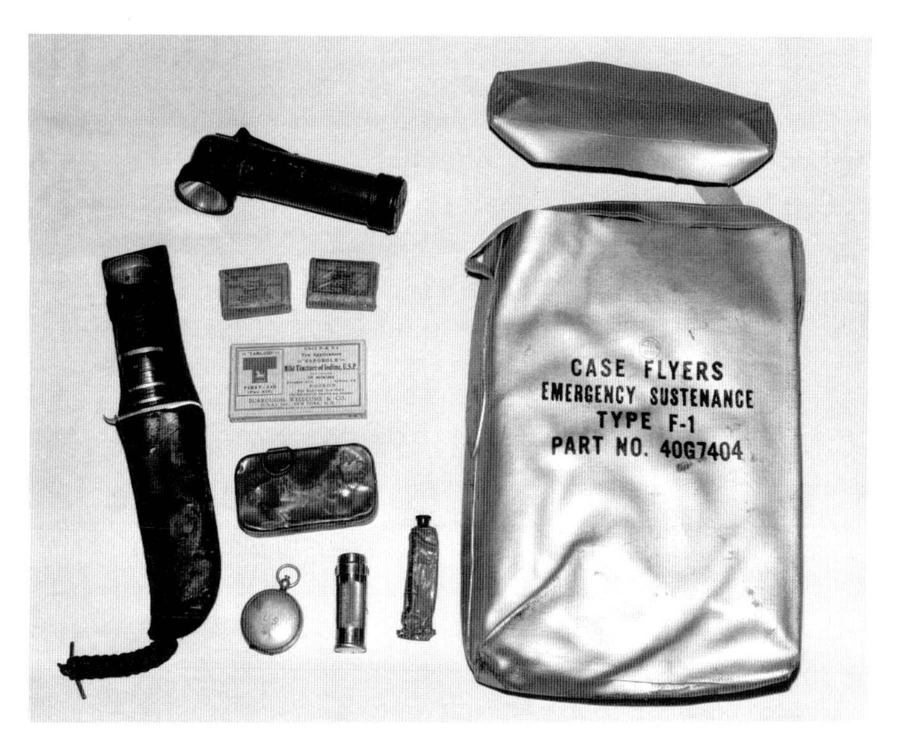

F-1 case showing partial components (inner metal container and rations missing)

The F-1 Emergency Sustenance Flyers Case is comprised of a 13" x 8" x 2" container made of aluminum painted rubberized canvas. All components were issued in a metal box inserted into the canvas container along with two tins of emergency rations. Very little information is available about the F-1 except that it was probably first used in the late 30's and early 40's.

Components for the F-1:

-Pocket compass
-Hunting knife
-Gauze bandages (2)
-Flashlight
-Carlisle bandage
-Iodine swabs
-Matches in waterproof match case (Marples)
-Fishing line
-Boric acid

E-3 AND E-3A 'PERSONAL AIDS' KITS

WWII 'personal aids' kits (left to right): E-3, U.S.N. E-3A, A.A.F. E-3A

The Army Air Force E-3 'personal aids' survival kit was used in World War II and consists of a 4 1/8" x 5 3/8" x 1 5/16" clear plastic container that was carred in a cloth bag. Tape was used to secure the two halves of the container.

The nomenclature was molded into the lid:

KIT: EMERGENCY SUSTENANCE
TYPE E-3
SPECIFICATION NO. 04-40441
QUANTITY 20,000
STOCK NO. 3300-550138
ORDER NO. 43-20112AF
NATHAN PRODUCTS CORP.
PROPERTY
AIR FORCES, U.S.ARMY

Components for the E-3:

-Safety matches
-Pocket compass
-6" hack saw blade
-Halazone tabs (12)
-Benzedrine tabs (6)
-U.S. Army field ration 'D' bar
*** -Package of 15 flavored dextrose tabs (6 lemon, 9 malted milk)**
*** -Envelope of instant bouillon powder (20 grams)**
*** -Chewing gum (1 stick)**

*** U. S. Army Bailout Ration was also used and contained these items.**

E-3 AND E-3A 'PERSONAL AIDS' KITS

(left): early A.A.F. E-3A (using tape to secure lid), later A.A.F. E-3A showing reverse side, match case & button compass

The A.A.F. and Navy E-3A 'personal aids' survival kits of World War II use identical containers and differ only in their contents. Both use 6" x 4 1/4" x 1 1/4" transparent ethlyl cellulose flasks designed to fit in a shirt pocket. The lid of the flask has a threaded cap and can be used to carry water when the contents have been removed. The lid is secured by a wire bail. Earlier containers do not have the bail wire and use tape to secure the lid.

The directions for the using the components are printed on the outside of the flask.

Components for the A.A.F. E-3A:

-Plastic match case w/matches
-Bouillon powder (2 pkgs.)
-2 oz. bars sweet chocolate (2)
-1 box caramels
-Chewing gum (4 sticks)
-Antiseptic ointment
-Benzedrine sulfate tabs
-Halazone tabs
-Aspirin tabs
-4" hack saw blade
-Small compass
-Adhesive tape

Components for the Navy E-3A:

-4" hack saw blade
-Matches in plastic match case
-Adhesive tape
-Atabrine (for malaria)
-Sulfaguanidine (for diarrhea)
-Antiseptic ointment
-Halazone tablets
-Aspirin
-Benzedrine sulfate (to prevent sleep)
-Bouillon powder
-Emergency ration
-Chewing gum (4 sticks)

E-17 'PERSONAL AIDS' KIT

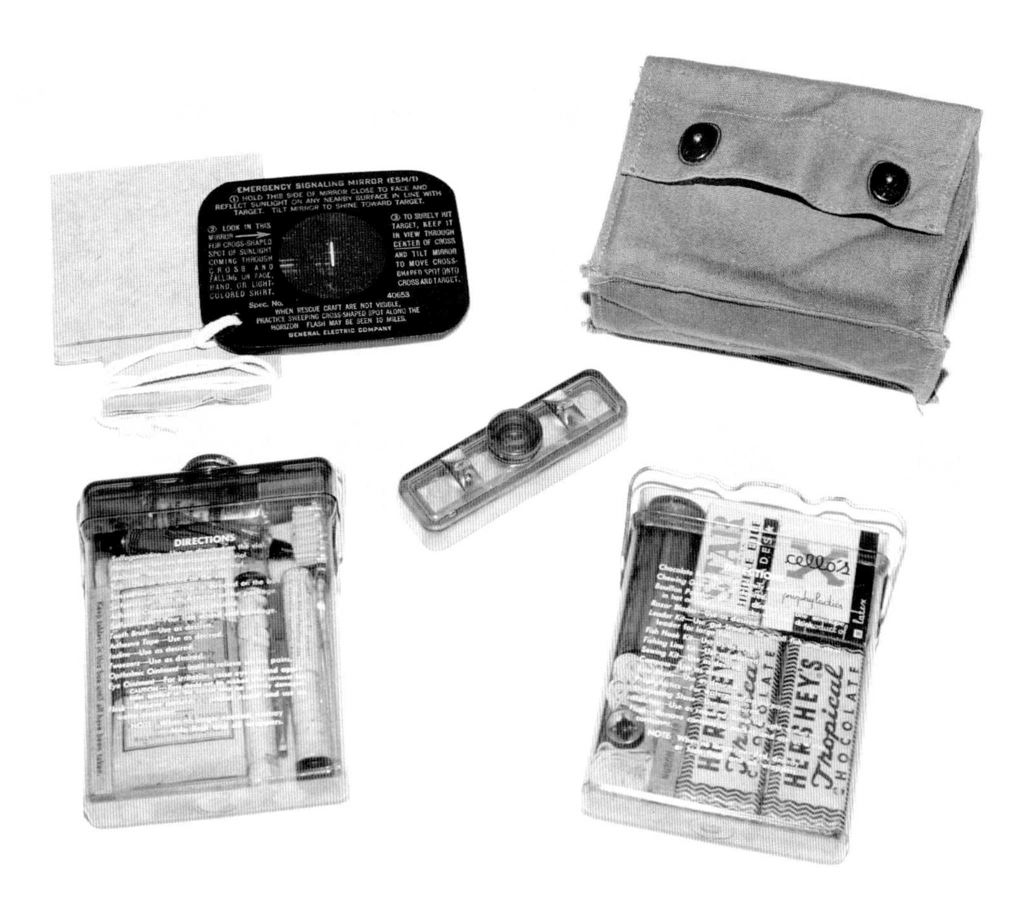

E-17 'personal aids' kit showing ESM-1 mirror and protective cardboard enclosure, canvas carrying case, medical flask and general flask (Lid removed from general flask to show bail wire.)

The E-17 'personal aids' kit was first used during World War II and was still listed in Air Force supply catalogs in the early 1950's. Two transparent ethyl cellulose flasks each measuring 6" x 4 1/4" x 1 1/4" and an ESM-1 mirror are enclosed in a cotton duck container that could be hung from a regular or standard web pistol belt. The lid of each flask was sealed by a rubber gasket and retained by a wire bail. Each lid also had a small threaded, gasketed cap that permitted the flask to be used as a water container after all the contents were removed. One flask contained medical items and the other had specific sustenance and survival items.

Components in the medical flask:

-Adhesive compress (6)
-Adhesive tape (1 yard)
-Atabrine tablets (1 vial)
-Benzedrine sulfate tablets (6)
-Eye ointment (2 tubes)
-Halazone tablets (1 vial)
-Opthalmic ointment (1 tube)
-Salt tablets (2 vials)
-Sulfadiazine tablets (8)
-Sulfanilamide, sterilized (2 env.)
-Sulfaguanidine tablets (4 vials)
-Tincture of iodine (1 bottle)
-Tweezers (1 pair)

E-17 'PERSONAL AIDS' KIT

Components in the general flask:

- -Match case with matches (1)
- -Tooth brush (1)
- -Brass compass (1)
- -Chewing gum (4 sticks)
- -Hack saw blade (1)
- -Kit, fish hook (1)
- -Kit, fish line (1)
- -Kit, leader (1)
- -Kit, sewing (1)
- -Bouillon powder (2)
- -Prophylactic (3) (Used as water containers)
- -Chocolate rations (4) ('Hershey bars' with red labels)
- -Razor blades (10)
- -Sharpening stone (1)

PSK-2 PERSONAL SURVIVAL KIT

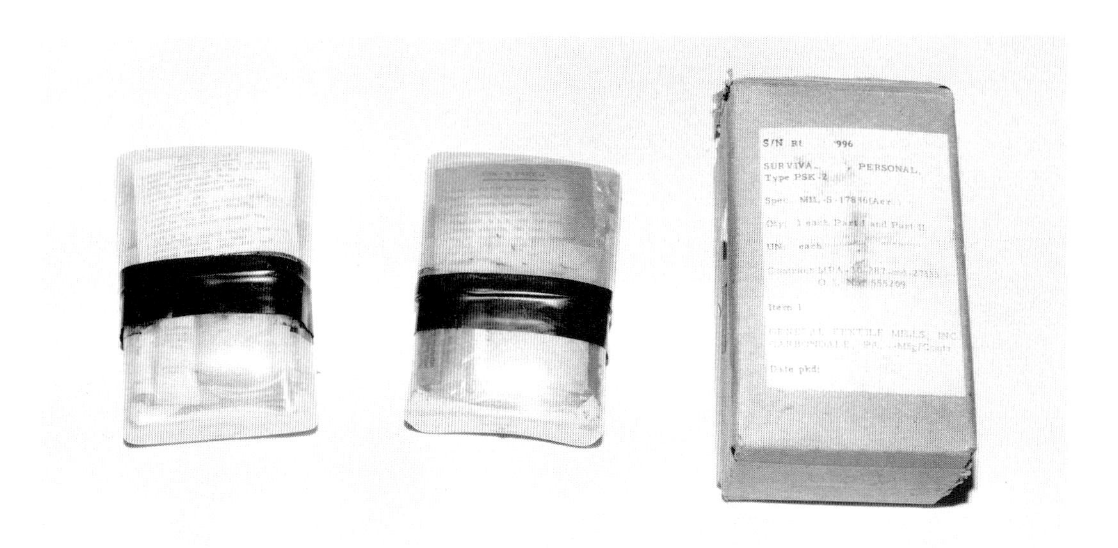

PSK-2 Personal Survival Kit

The PSK-2 Personal Survival Kit was used by the U. S. Navy in the 1950's and 60's and consisted of two clear polyethelene tape-sealed containers each measuring 4 1/2" x 3 1/4" x 1 1/2". It is designed to fit into the pockets of the flight suit.

Components for the PSK-2:

PART I

-Adhesive plaster (1)
-Book matches (1)
-Compress, guaze (1)
-Bouillon cubes (1 pkg.)
-Terramycin tablets (1 vial)
-Chocolate ration bar (1)
-D'Amphetamine sulfate tablets (1 pkg.)
-APC tablets (1 vial)
-Tetracaine ophthalmic ointment (1 tube)
-Chloroquine phosphate tablets (1 vial)
-Bandaids (5)
-Aluminum foil (1 piece)
-Soaped tissue (1 pkg.)
-Sewing kit (1)

PART II

-Adhesive plaster (1)
-Book matches (1)
-Compress, guaze (1)
-Bouillon cubes (1 pkg.)
-Bandaids (5)
-Soaped tissues (1 pkg.)
-Water purification tablets (1 bottle)
-Benzalkonium chloride tincture (bottle)
-Petrolatum tube (1)
-Hacksaw blade (1)
-Razor blade (1)
-Chapstick (1 tube)

1 AND 2-PART INDIVIDUAL SURVIVAL KITS

The 1 and 2-Part Individual Survival Kits are similar to the Navy PSK-2 kit. The containers are slightly larger, measuring 4 1/2" x 3 1/2" x 1 3/4" and components vary depending upon the year of issue. The 2-part kits were issued in the early 1960's and may have seen some service in the late 50's. The 1-Part Individual Survival Kit is still being issued today.

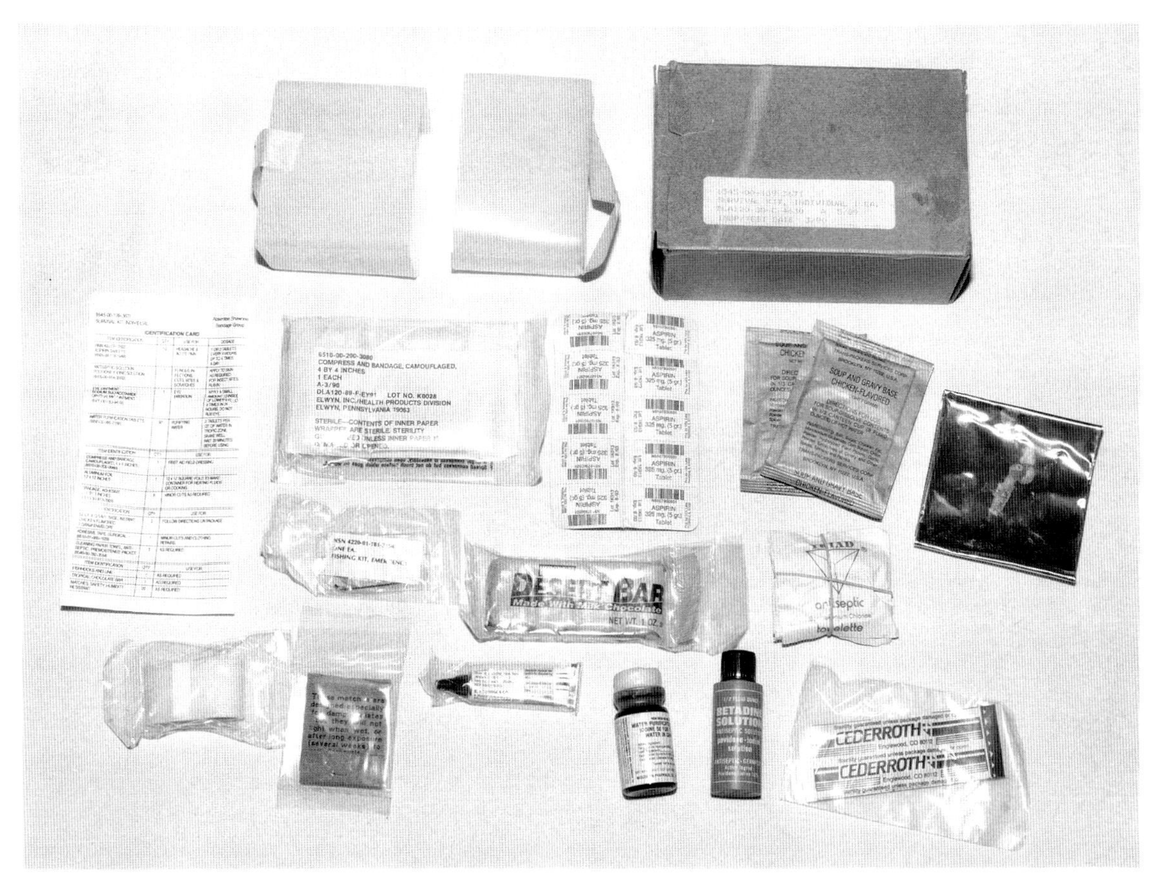

1-Part Individual Survival Kit components (1989)

Components for the 1989 dated 1-Part Individual Survival Kit (NSN6545-00-139-3671):

ITEM	STOCK NUMBER
-Pain killer tabs, aspirin tablets (10)	6505-00-118-1948
-Antiseptic solution, povidone iodine solution (1)	6505-00-914-3593
-Eye Ointment, Sodium Sulfacetamide Ophthalmic Ointment (1)	6505-00-183-9419
-Water purification tablets (50)	6850-00-965-7166
-Compress and bandage, camouflaged, 4" x 4" (1)	6510-00-200-3080
-Aluminum Foil, 12" x 12" (1)	
-Bandage, adhesive 3/4" x 3" (6)	6510-00-913-7909
-Soup & gravy base, instant, chicken flavored, 7 gram envelope (2)	
-Adhesive tape, surgical (1)	6510-01-060-1639
-Cleaning paper towel, antiseptic, premoistened packet (3)	8540-00-782-3554
-Fish hooks and line (1)	
-Tropical chocolate bar (1)	
-Matches, safety, humidity resistant (20)	

1 AND 2-PART INDIVIDUAL SURVIVAL KITS

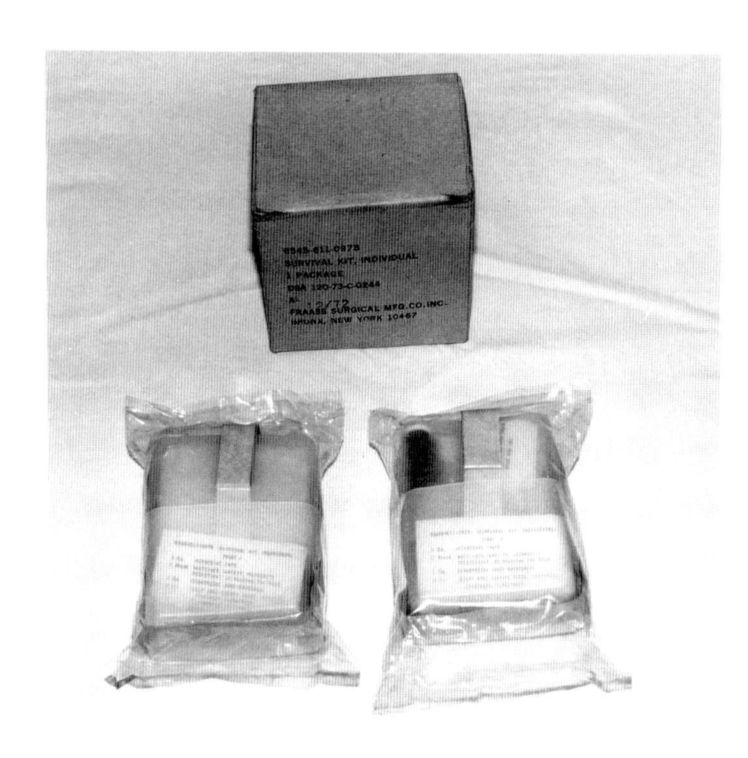

2-Part Individual Survival Kit showing each kit
in its sealed clear plastic baggie (1972)

Components for the 1972 dated 2-part kit (FSN 6545-611-0978):

PART I

1 Ea.	ADHESIVE TAPE
1 Book	MATCHES, SAFETY, HUMIDITY RESISTANT, 20 Matches Per Book
1 Ea.	COMPRESS, AND BANDAGE
2 Env.	SOUP AND GRAVY BASE, INSTANT, CHICKEN, FLAVORED (1 envelope in 1 cupful of hot water)
1 Ea.	SWEET CHOCOLATE RATION BAR (CONSERVE)
1 Pkg.	ASPIRIN TABLETS, 12s
1 Tube	SODIUM SULFACETAMIDE OPHTHALMIC OINTMENT, MODIFIED, 1/8 OZ.
1 Strip	CHLOROQUINE and PRIMAQUINE PHOSPHATES TABLETS (See Directions on reverse side)
1 Pkg.	BANDAGE, ADHESIVE
1 Ea.	ALUMINUM FOIL Directions: Fold to make container for heating fluids or cooking
1 Pkg.	SOAPED TISSUES, WRAPPED (Dampen skin, rub with tissue, rinse, and dry)
1 Ea.	SEWING KIT
1 Pkg.	FISH HOOKS AND LINE

PART II

1 Ea.	ADHESIVE TAPE
2 Book	MATCHES, SAFETY, HUMIDITY RESISTANT, 20 Matches Per Book
1 Ea.	COMPRESS, AND BANDAGE
2 Env.	SOUP AND GRAVY BASE, INSTANT, CHICKEN, FLAVORED (1 envelope in 1 cupful of hot water)
1 Ea.	SWEET CHOCOLATE RATION BAR (CONSERVE)
1 Pkg.	BANDAGE ADHESIVE
1 Pkg.	SOAPED TISSUES, WRAPPED (Dampen skin, rub with tissue, rinse and dry)
1 Bottle	WATER PURIFICATION TABLETS, IODINE
1 Bottle	PROVIDONE - IODINE SOLUTION (Skin Disinfectant)
1 Tube	PRETROLATUM
1 Ea.	FLEXIBLE SAW
1 Ea.	RAZOR, SURGICAL PREPARATION, STRAIGHT TYPE
1 Tube	LIPSTICK, ANTI-CHAP, HOT CLIMATE 1/8 OZ.

1 AND 2-PART INDIVIDUAL SURVIVAL KITS

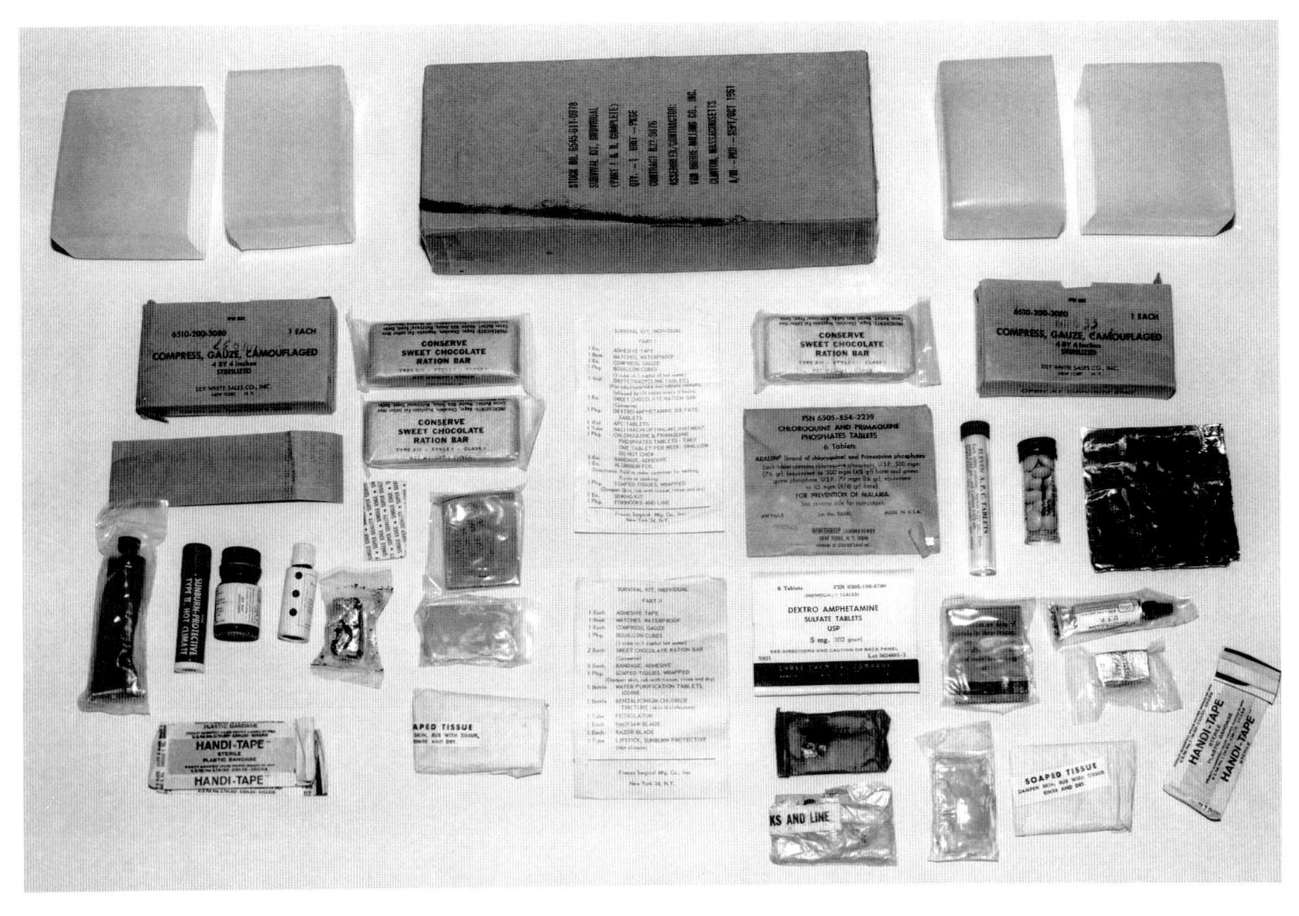

2-Part Individual Survival Kit components (1961)

Components for the 1961 dated 2-part kit (P/N 6545-611-8978):

PART I

1 Ea.	ADHESIVE TAPE
1 Book	MATCHES, WATERPROOF
1 Ea.	COMPRESS, GAUZE
1 Pkg.	BOUILLON CUBES (1 cube in 1 cupful of hot water)
1 Vial	OXYTETRACYCLINE TABLETS (For infections take two tablets initially followed by (1) tablet every 6 hours.)
1 Ea.	SWEET CHOCOLATE RATION BAR (Conserve)
1 Pkg.	DEXTRO-AMPHETAMINE SULFATE TABLETS
1 Vial	APC TABLETS
1 Tube	BACITRACIN OPHTHALMIC OINTMENT
1 Pkg.	CHLOROQUINE & PRIMAQUINE PHOSPHATES TABLETS - TAKE ONE TABLET PER WEEK, SWALLOW- DO NOT CHEW
1 Each	ALUMINUM FOIL Directions: Fold to make container for heating fluids or cooking.
1 Pkg.	SOAPED TISSUES, WRAPPED (Dampen skin, rub with tissue, rinse and dry.)
1 Ea.	SEWING KIT
1 Pkg.	FISH HOOKS AND LINE

PART II

1 Ea.	ADHESIVE TAPE
1 Book	MATCHES WATERPROOF
1 Each	COMPRESS, GAUZE
1 Pkg.	BOUILLON CUBES (1 cube in 1 cupful hot water)
2 Each	SWEET CHOCOLATE RATION BAR (Conserve)
5 Each	BANDAGE, ADHESIVE
1 Pkg.	SOAPED TISSUES, WRAPPED (Dampen skin, rub with tissue, rinse and dry.)
1 Bottle	WATER PURIFICATIONTABLETS, IODINE
1 Bottle	BENZALKONIUM CHLORIDE TINCTURE (skin disinfectant)
1 Tube	PETROLATUM
1 Each	HACKSAW BLADE
1 Each	RAZOR BLADE
1 Tube	LIPSTICK, SUNBURN PROTECTIVE (Hot climate)

1 AND 2-PART INDIVIDUAL SURVIVAL KITS

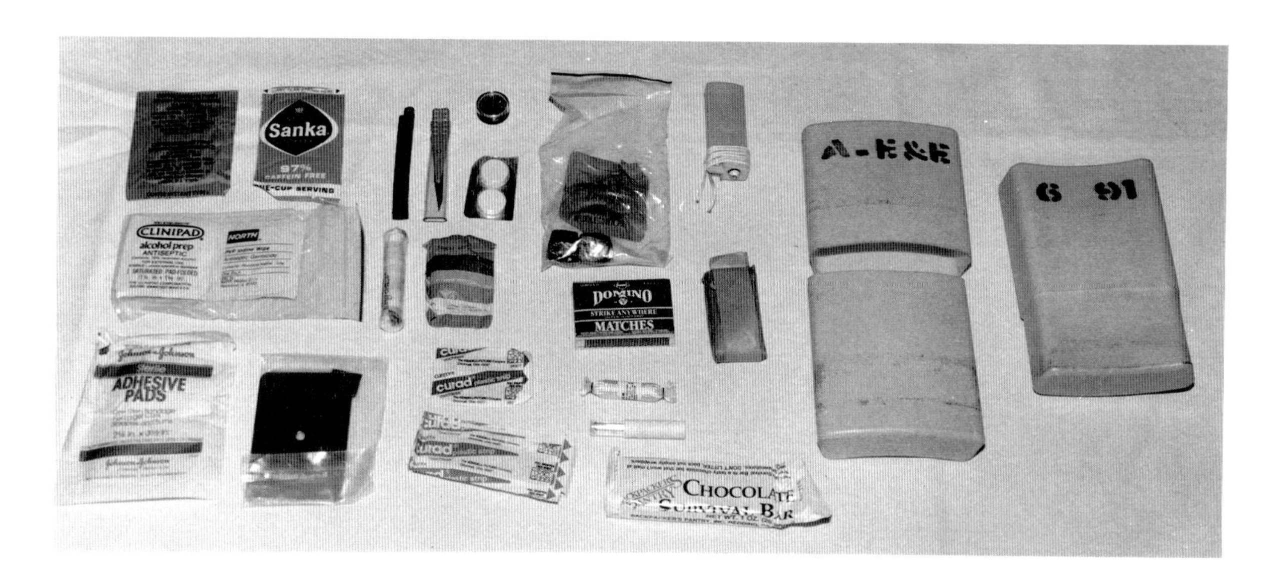

Special E & E Kit (1991) components (top row, left to right): sugar, coffee, hacksaw blade, surgical razor, razor, button compass, pain pills, salt and bouillon cubes in ziplock bag, flashlight (middle row, left to right): antiseptic pad, candies (2), matches, toilet issue (bottom row, left to right): adhesive pads, 2" x 3" metal mirror, small and large bandaids, ammonia inhalents, iodine capsule, chocolate bar

The origin and use of this 1-Part E & E Kit are unknown but may have been put together and used during Desert Storm. The 2-piece plastic container is solid yellow (not clear plastic as in other 1 and 2-part kits) and has no markings other than that shown in the photo. The reverse side of another kit is shown indicating the date.

SEEK 1 KIT

SEEK 1 Kit

The SEEK 1 (Survival, Escape and Evasion Kit) was used by the Navy in the 1950's and 60's. Because of its availability, it may have been used by the other services as well. The SEEK 1 is comprised of two separate clear rigid plastic containers, each measuring 5 3/8" x 4 3/8" x 2". The lid on each container is secured by two aluminum channels and sealed by a rubber gasket. A vent port is also included on each container.

Components are listed on each lid.

<u>**LABELS**</u>

(on bottom of each container)

SURVIVAL,ESCAPE,AND EVASION KIT,
INDIVIDUAL AIRMEN'S, TYPE SEEK-1
CONTAINER NO. 1
Specification XRAAE-157
Second Quarter 1964
Lite Industries, Inc.
Contract No. N383-84452A

SURVIVAL,ESCAPE, AND EVASION KIT,
INDIVIDUAL AIRMEN'S, TYPE SEEK-1
CONTAINER NO. 2
Specification XRAAE-157
Second Quarter 1964
Lite Industries, Inc.
Contract No. N383-84452A

The shipping box lists only one FSN for both containers: RM4240-731-9909-LA20

SEEK 1 KIT

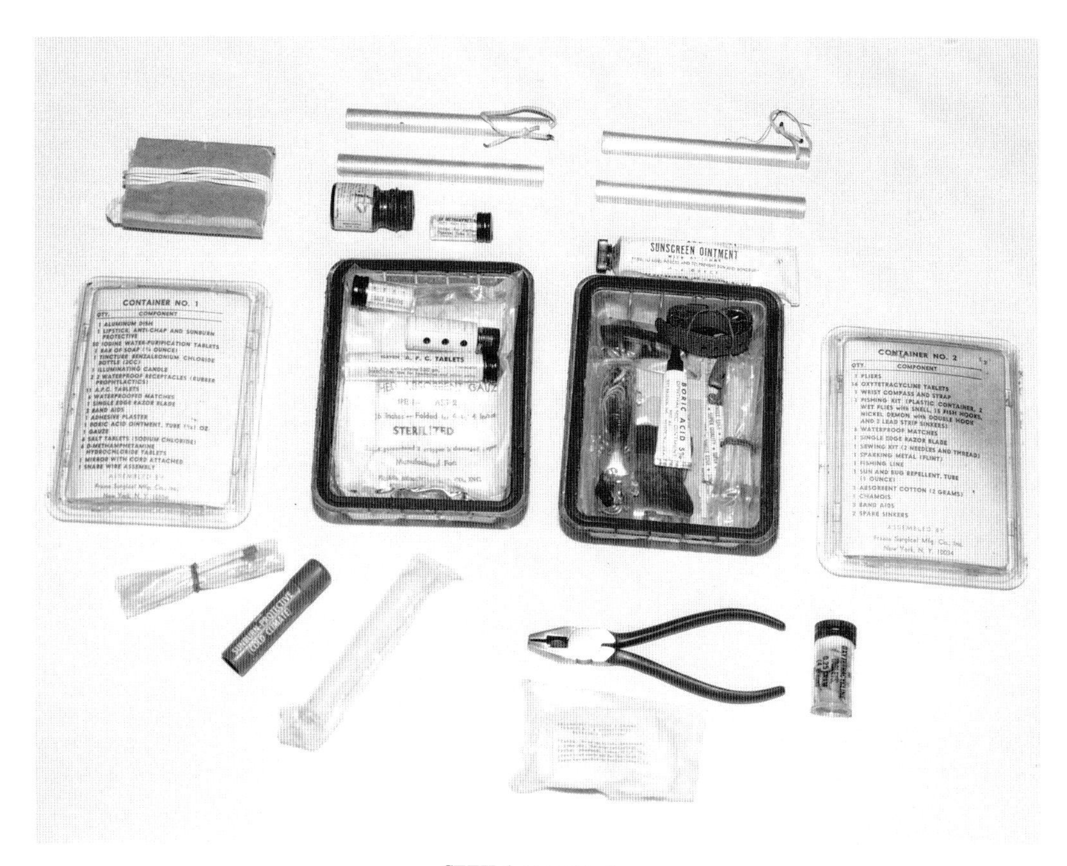

SEEK 1 components

SEEK 2, SRU-31/P, AIRMAN'S INDIVIDUAL SURVIVAL KITS

The SEEK 2, Airman's 24 hours SRU-31/P and the Airman's Individual Survival Kits have 2 parts: a Medical and General packet. These are components of various survival vests and were first used in Vietnam as part of the Navy's SV-1, SV-2 and the Army's OV-1 vests. These are all very similar in design. The Airman's Individual Kit and SRU-31/P are still being used today.

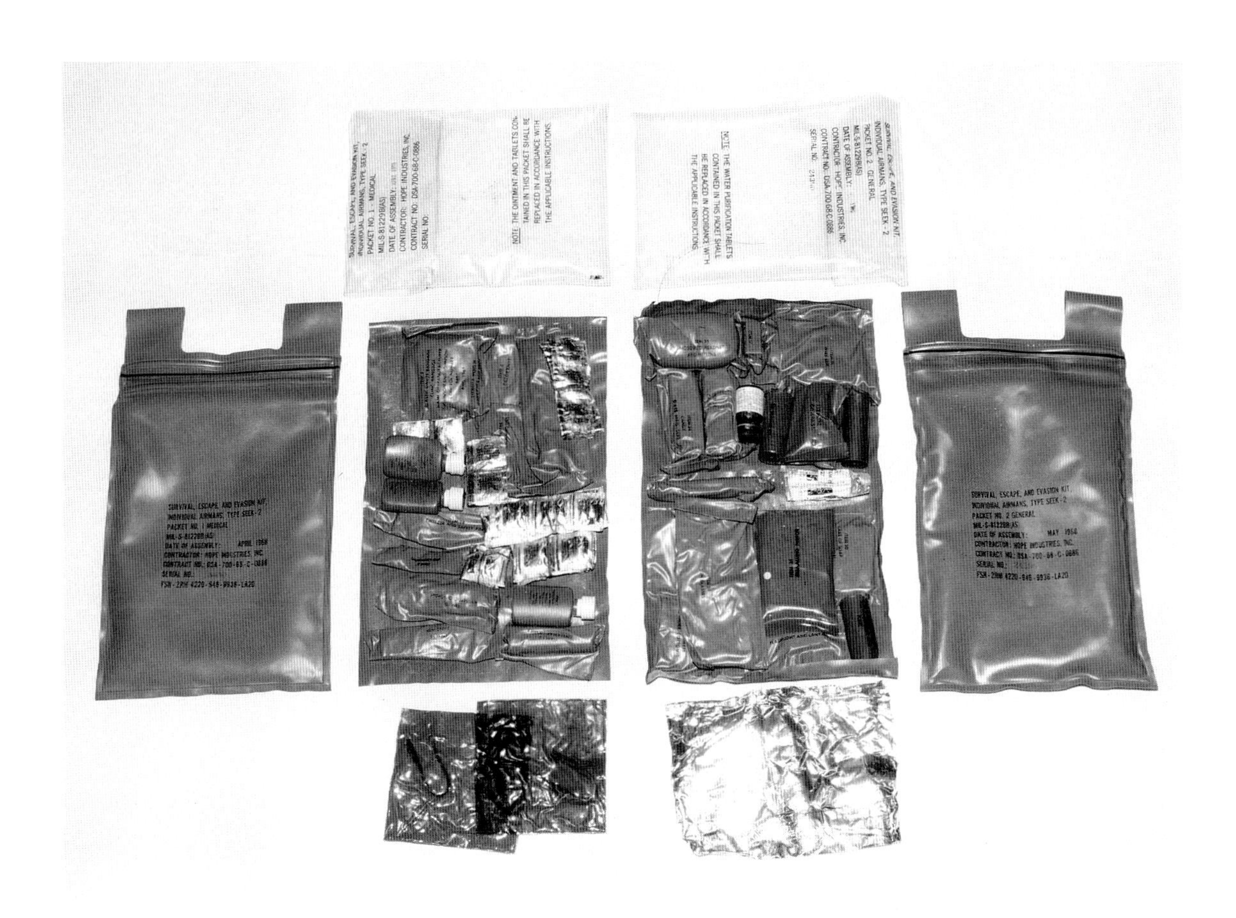

SEEK 2 Kit (top row): inner sealed bags (middle row, left to right): medical outer bag, medical components general components, general outer bag (bottom row): red and blue strobe covers, aluminum foil

The SEEK 2 (Survival, Escape and Evasion Kit) was the initial design and employed green poly outer bags that were sealed with a ziplock type closure. The 13 1/2" x 7 1/2" outer bags contained a sealed clear poly inner bag with the indiviual components (sealed in their own green poly 'labeled' bags) attached to a 13 3/4" x 7 7/8" (opened) folded plastic sheet by a tacky 'glue'. The outside of the folder had the component positions printed in place.

SEEK 2, SRU-31/P, AIRMAN'S INDIVIDUAL SURVIVAL KITS

Components for the SEEK 2:

MEDICAL
(FSN RM 4220-946-9936-LA20)

-Bandaids
-Antiseptic ointment
-Antidiarrhea
-Antimalaria
-Alert tablets
-Salt tablets
-Tweezers and safety pins
-Anti chap lipstick
-Anti infection
-Anti motion sickness
-Eye Ointment
-Liquid soap
-Leech repellent
-Insect repellent
-Elastic gauze bandage
-Burning lens
-Sun and bug repellent
-Razor knife
-Fungicidal powder
-Aspirin tablets
-Aluminum foil

GENERAL
(FSN Same as medical)

-Wire saw
-Candle
-Fishing kit
-Fishing line
-Wrist compass
-Water purification tablets
-Bouillon cubes
-Sewing kit
-Candy
-Mosquito headnet and mittens
-Flashlight and lanyard
-Cake of soap
-Sponge
-Waterproof receptacle
-Sunglasses
-Matches and flint
-Combo. hacksaw and knife blade
-Arrowhead
-Fire starter and tinder
-Blue and red filter for the strobe light

The SRU-31/P and the Airman's Individual Kits used Velcro for closure and contain fewer components. The green vinyl 12 5/8" x 6 3/4" outer pouches each contained a green vinyl inner pouch with two pockets, one for the standard components (Medical or General) and a pocket for optional components. The 10 3/4" x 4 3/4" medical and the 14 1/2" x 4 3/4" general folders in the inner pouch use hook and pile to secure the individual items and the optional pocket has an empty folder with hook tabs to attach any miscellaneous items.

SRU 31/P Medical and General kit (1972)

SEEK 2, SRU-31/P, AIRMAN'S INDIVIDUAL SURVIVAL KITS

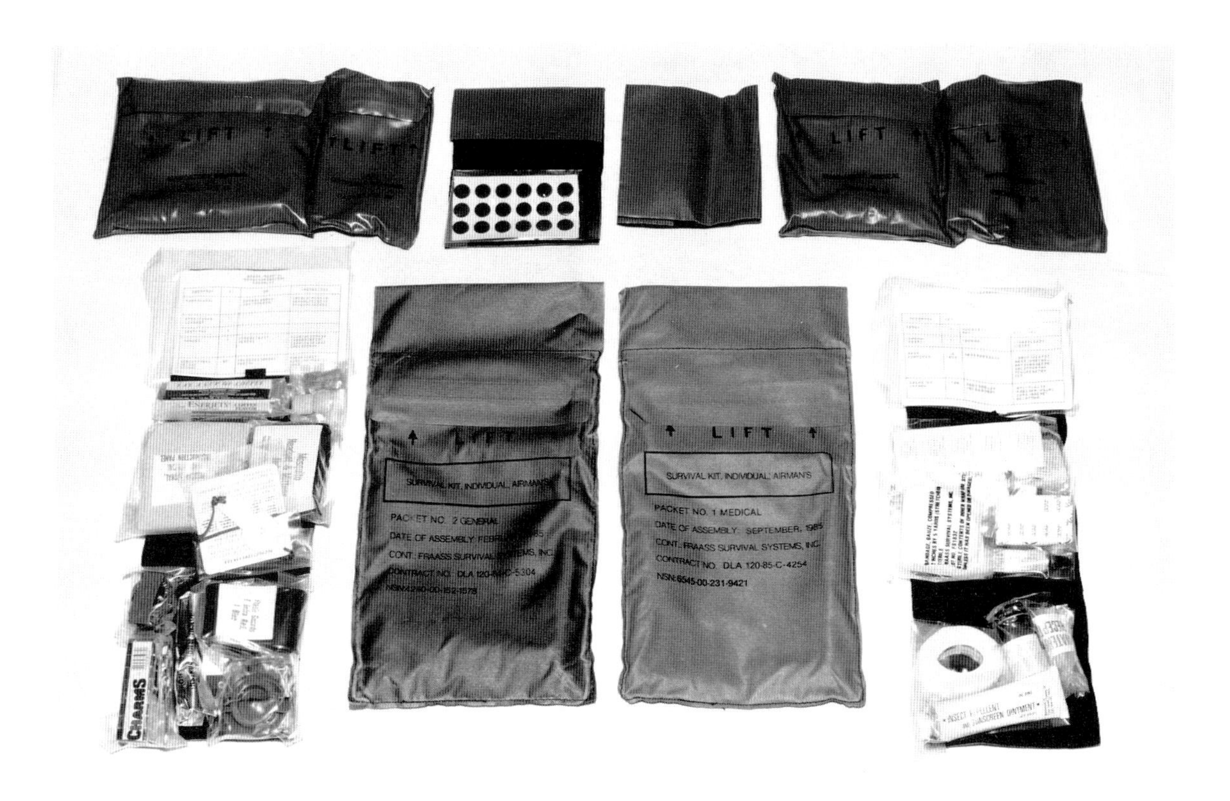

Individual Airman's Survival Kit (top row, left to right): general inner pouch, spare attachment sheet and Velcro 'dots' for General kit, spare attachment sheet for Medical kit, medical inner pouch (bottom row, left to right): General kit components, general outer bag, medical outer bag, Medical kit components

Components for the SRU-31/P and Airman's kits:

MEDICAL
SRU-31/P (FSN RD4240-231-9421-LX1X)
Airman's (NSN 6545-00-231-9421) (1974)

-Instruction card
-Soap
-Surgical tape
-Water receptacle
-Insect repellent
-Pain kiler
-Anti-diarrhea
-Bandaid
-Bacitracin (eye ointment)
-Bandage (elastic)
- Water purification tablets

GENERAL
SRU-31/P (FSN RD4240-152-1578-LX1X)
Airman's (NSN 6545-00-478-6504) (1974)

-Chiclets
-Tinder
-Compass (wrist)
-Water bag (1 quart)
-Signal panel
-Mirror
-Razor knife
-Tweezer and pins
-Mosquito headnet and mittens
-Charms
-Enerjets
-Metal match
-Flashguards (red and blue)

The only difference between components in the SRU-31/P and the Airman's Individual Kits is the replacement of pain killers with aspirin in the Airman's kit.

'TAC' KITS

The Individual Tropical Survival Kit, Tactical Air Crew (TAC kit) has been in use since the 1960's and is still being issued as part of the SRU-21/P Survival Vest. Components are currently carried in a clear ziplock bag. Earlier TAC kits consisted of a 4 3/4" x 3 1/4" x 2 1/8" two piece green or o.d. semihard plastic container shaped so that when the cover is removed, it folds out flat for ease in reaching contents. The contents are held into the inner part of the container by a 'tacky' glue to prevent loss. These kits have also been used with the SRU-19/P chaps, SRU-32/P vest and have been listed as a substitute first aid kit in various 'soft' and 'hard' seat kits.

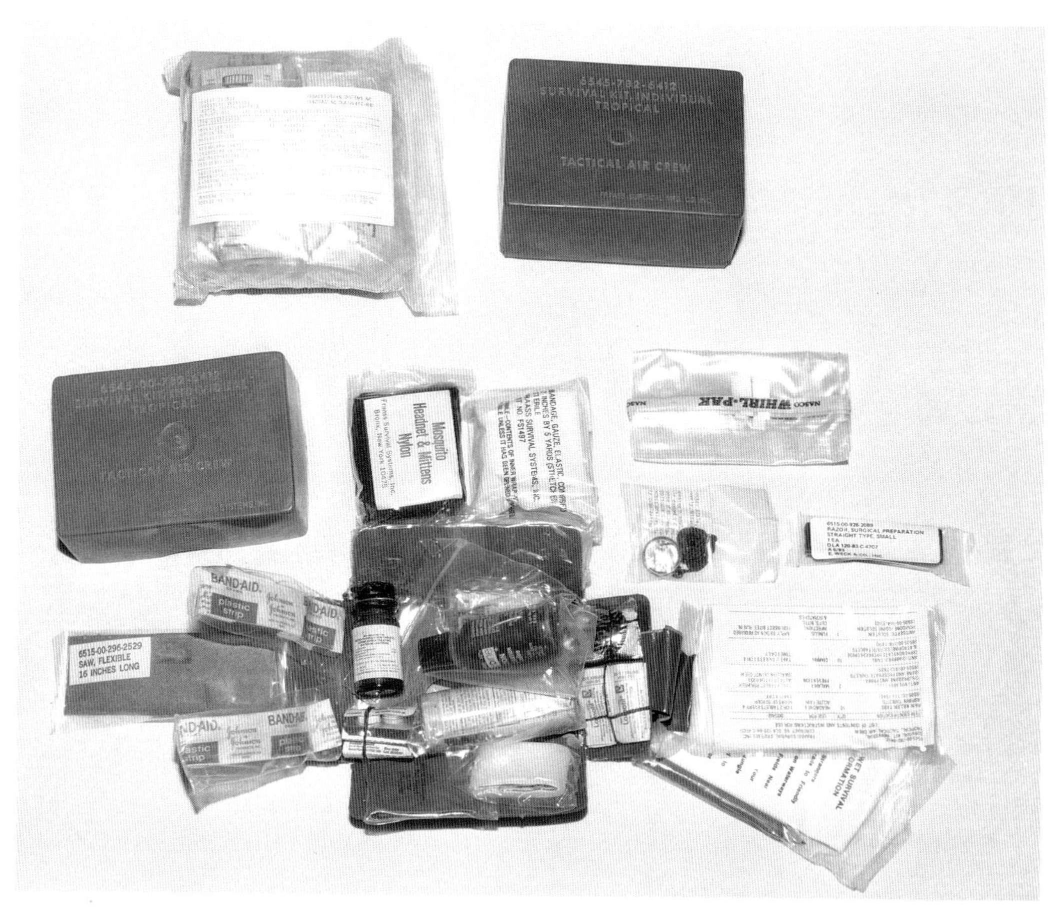

TAC kits (top row, left to right): current issue in ziplock bag, early issue (bottom row, left to right): 1980's issue showing lid removed and components on unfolded inner container (Very early kits were void of 'Tactical Air Crew' on the lid and listed a contract number in the lower right corner.)

'TAC' KITS

Components for the TAC kits:

FSN 6545-782-6412 (Early)

-Doxycycline Hyclate caps (anti-infection)
-Salt tabs
-Safety pins
-Needle and thread
-Dextroamphetamine sulfate tabs (stay awake)
-Propoxyphene hydrochloride, aspirin, caffeine, and phenacetin caps (pain killers)
-Stick of insect repellent
-Compass
-16" flexible saw
-Safety matches, waterproof
-Survival booklet and pencil
-Chloroquine and primaquine phosphate tabs (antimalarial)
-Diphenoxylate hydrochloride andatropine sulfate (antidiarrheal)
-Iodine
-Mosquito headnet and gloves
-Bandaids
-Water purification tabs
-Surgical razor
-Gauze bandage
-Surgicial tape
-Opthalmic ointment

NSN 6545-00-782-6412 (Later)

-Stick of insect repellent
-Compass
-16" flexible saw
-Safety matches, waterproof
-Survival booklet and pencil
-Food sample bags (3)
-Aspirin tabs
-Antimalaria tabs
-Antidiarrhea tabs
-Iodine
-Mosquito headnet and mittens
-Bandaids
-Water purification tabs
-Surgical razor
-Gauze bandage
-Surgical tape (1 yd.)
-Opthalmic ointment

'TAC' KITS

Components for the TAC kits (continued):

NSN 6545-01-120-2632 (Current)

-Aspirin
-Water purification tabs
-Matches, waterproof
-Eye ointment
-Food bags (3)
-16" flexible saw
-Bandaids
-Small bar soap
-Antidiarrheal tabs
-Bandage gauze
-Mosquito repellent
-Antimalarial tabs
-Skin closures
-Betadine ointment (iodine)

HOT-WET ENVIRONMENT INDIVIDUAL SURVIVAL KIT

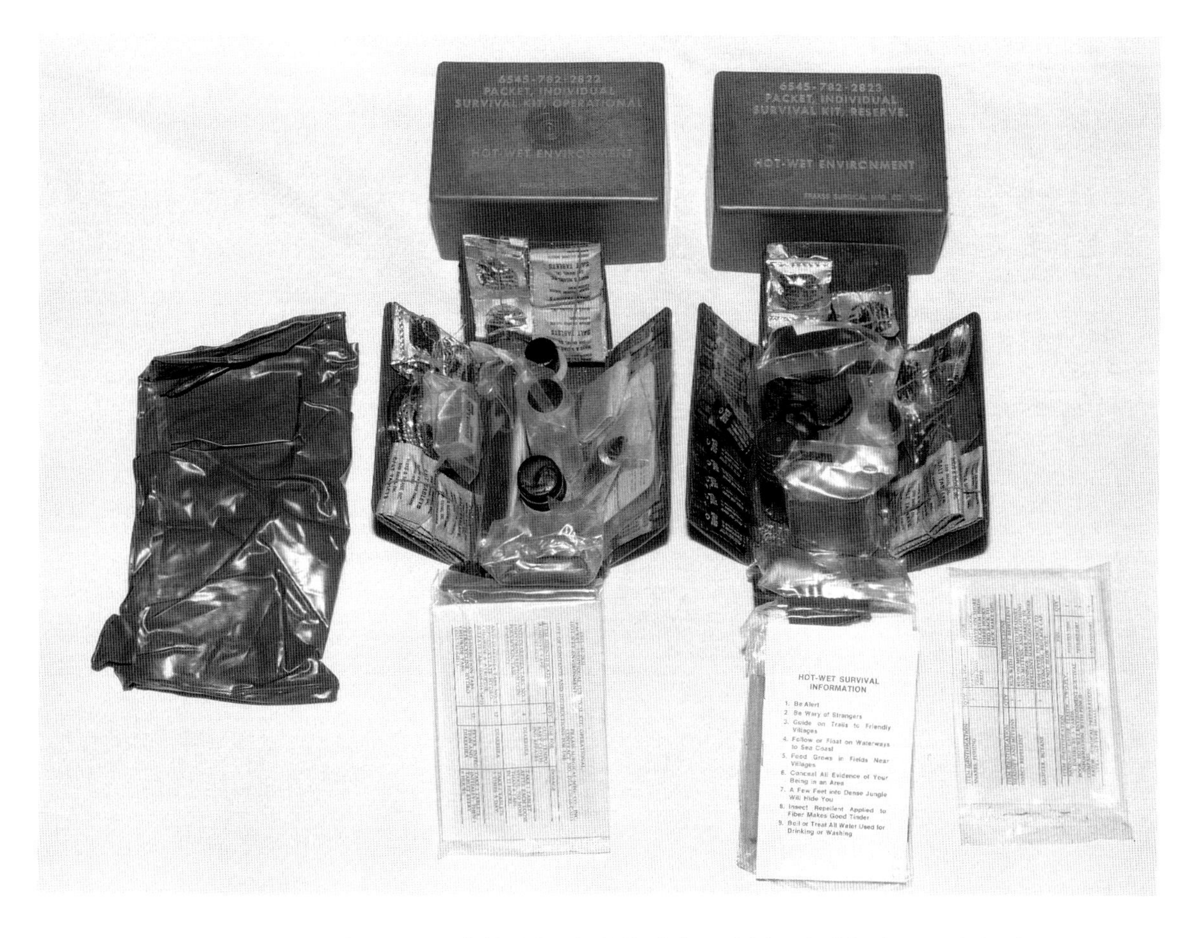

2-Part Hot-Wet Environment Individual Survival Kit (left to right): vinyl food storage/misc. bag Operational kit (lid removed to show components), Reserve kit (lid removed to show components)

The 2-Part Hot-Wet Environment Individual Survival Kit was used in Vietnam, primarily by special operations groups such as Army Special Forces and Rangers. It was not designated for use by aircrews but the availability could have lent its use to Army helicopter and scout aircrews. This kit used a similar container and shared many components from the TAC kits. The Hot-Wet kit used two separate containers each placed in a nylon carry bag. These bags used a Velcro fastener for belt attachment. Components were almost indentical to the earlier TAC kits and many items were doubled in quantity. Components unique to the Hot-Wet kit were a butane lighter, a 4" long pengun type flare launcher and two red flares, signal mirror, bouillon cubes, fishing line and fishhooks and 2 small vinyl food/equipment bags. The two containers were designated:

6545-782-2822 PACKET, INDIVIDUAL SURVIVAL KIT, OPERATIONAL
6545-782-2823 PACKET, INDIVIDUAL SURVIVAL KIT, RESERVE

Other variations of nomenclature exist and some early kits had the contents printed on the outside of the container.

HOT-WET ENVIRONMENT
INDIVIDUAL SURVIVAL KIT

Operational and Reserve kits showing nylon carrying bag and packing box

Components for the Operational kit:

- -Pengun flare launcher and 2 flares
- -Finger saw
- -Signal mirror
- -Eye ointment
- -Fungicidal ointment
- -Water purification tabs
- -Antidiarrheal tabs
- -Bouillon cubes
- -Betadine solution
- -Adhesive tape
- -Salt tabs
- -Insect repellent
- -Bandaids
- -Gauze bandage
- -Instructions

Components for the Reserve kit:

- -Providine-iodine ointment
- -Butane lighter
- -Water purification tabs
- -Antidiarrheal tabs
- -Antimalarial tabs
- -Bouillon cubes
- -Mosquito headnet and gloves
- -Insect repellent
- -Betadine solution
- -Adhesive tape
- -Salt tabs
- -Bandaids
- -Gauze bandage
- -Pencil and writing tablet
- -Instructions
- -Fishing line and hooks
- -Sewing kit
- -Compass
- -Razor knife
- -Fungicidal ointment
- -Pain killers
- -Dextroamphetamins (stay awake pills)

SRU-16/P MINIMUM SURVIVAL KIT

The SRU-16/P Minimum Survival Kit was introduced in 1962 and is located in a small pocket attached to a torso harness or inside a parachute pack. The SRU-16/P's, installed inside parachute packs, are normally not accessible for use unless the parachute has been opened and are considered a onetime use item. Kits manufactured from 1962 through 1971 consist of a fabric outer container, approximately 7 1/2" long and 1 1/2" wide, a moisture proof sealed film inner container, and a minimum number of survival items. The outer container is stitched closed to prevent loosing the items.

Kits manufactured in 1972 and 1973 were similar except for the addition of a fiberboard inner container within the moisture proof container and Velcro tape in lieu of stitching. The water purification tablets and bandaids are attached to the moisture proof intermediate container for ease of inspection.

Kits manufactured in 1974 returned to the stitching and deleted the water purification tablets and bandaids.

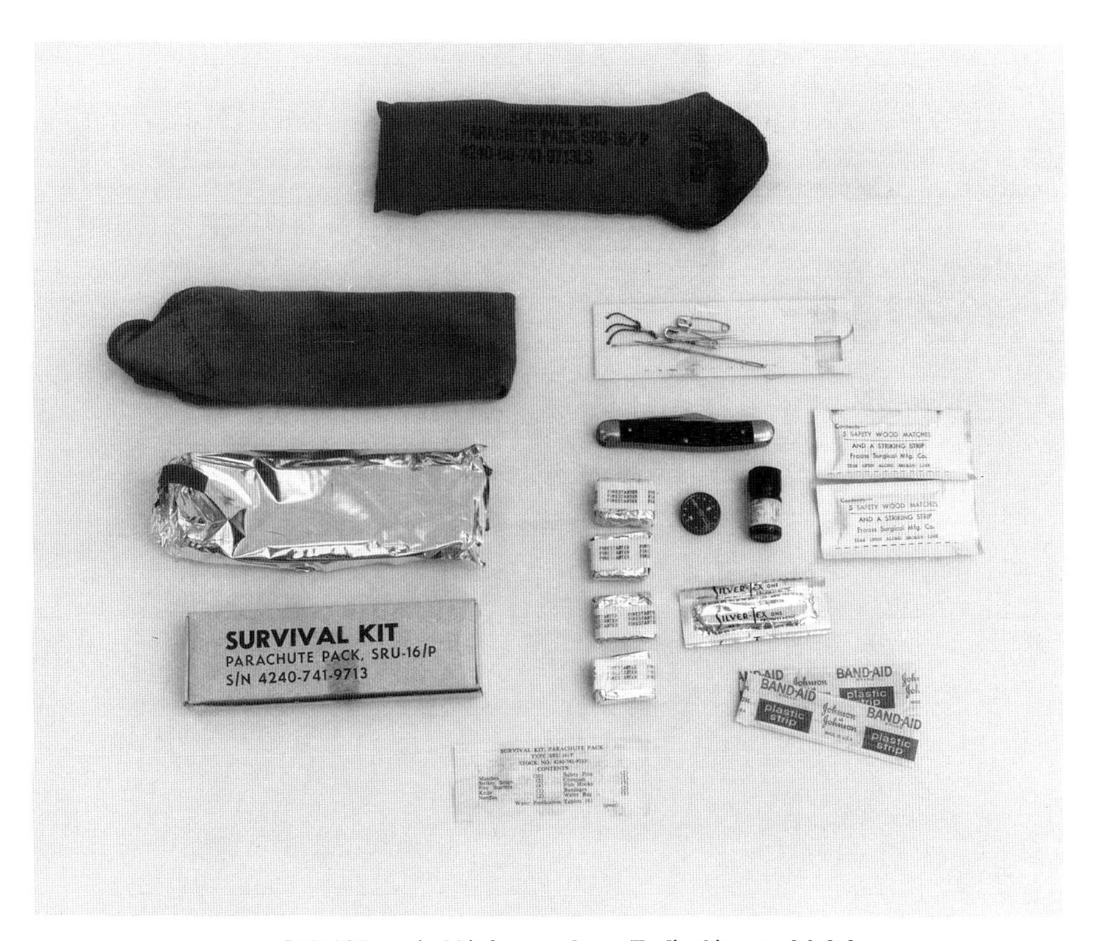

SRU-16/P survival kit for parachutes (Earlier kits were labeled with a large 'SURVIVAL' on the outer container.)

SRU-16/P MINIMUM SURVIVAL KIT

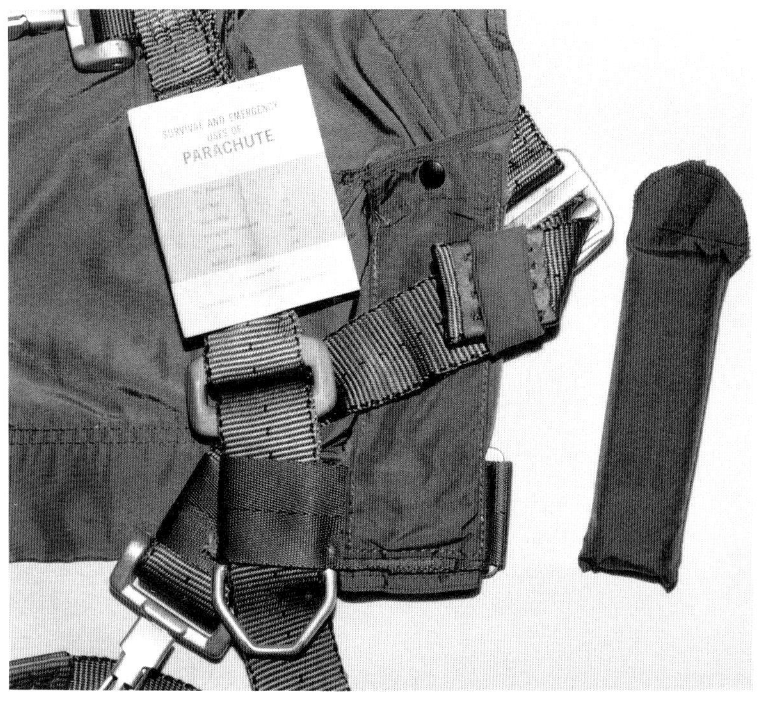

Above: SRU-16/P showing stowage pocket location on parachute torso harness ('Survival and Emergency Uses of Parachute' booklet also included in pocket.)

Right: Air Force parachute torso harness

Contents for 1962 through 1972 SRU-16/P's (FSN 4240-741-9713):

- -Matches (10)
- -Fire starters (3 or 4)
- -Striker strip (20)
- -Pocket knife (1)
- -Safety pins (2)
- -Water purification tabs (6)
- -Needles (2)
- -Compass, button (10)
- -Fish hooks (3)
- -Bandaids (2)
- -Water bag (1)
- -Instruction sheet (1)

Contents for later SRU-16/P's (NSN 4240-00-741-9713LS):

- -Compass button (1)
- -Fish hooks (3)
- -Knife (1)
- -Water bag (1)
- -Matches (10)
- -Needles (2)
- -Fire starters (4)
- -Stainless steel wire (20 feet)
- -Striker strip (2)
- -Instructions and contents sheet (1)

WORLD WAR II FIRST AID KITS, PERSONAL TYPE

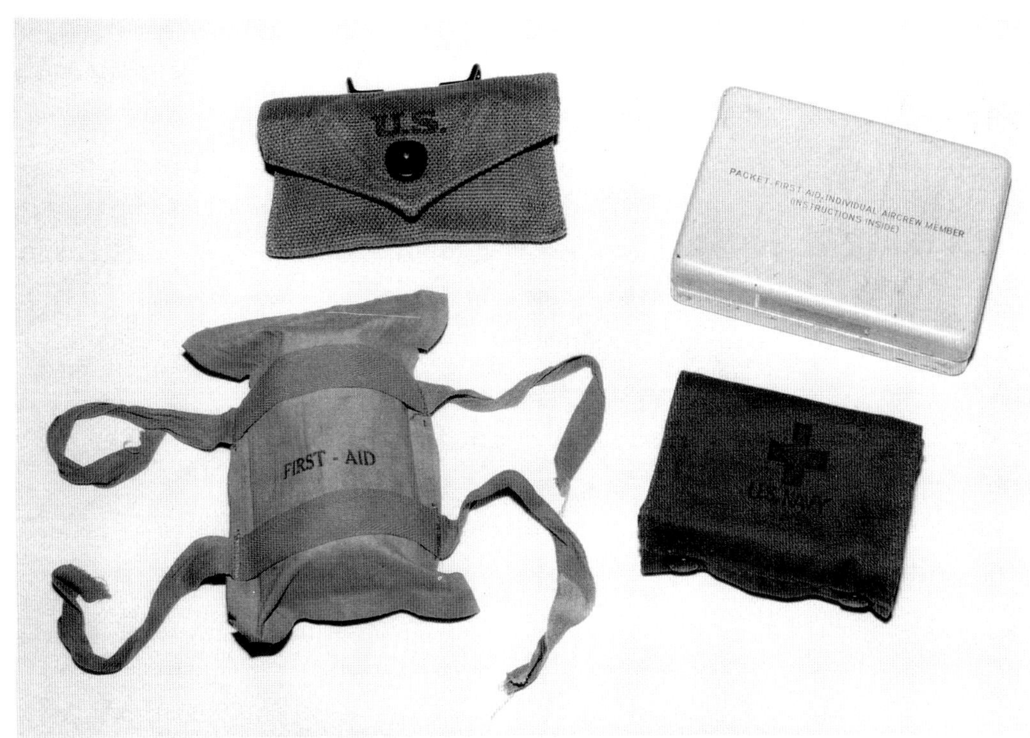

WWII individual first aid kits (top row, left to right): standard cloth first aid pouch, A.A.F. aluminum first aid packet (bottom row , left to right): parachute first aid packet, U.S.N. Aviators kit

WORLD WAR II U.S.N. AVIATOR'S KIT, INDIVIDUAL

This kit consists of a 4 1/2" x 2" x 3 1/8" outer herringbone-twill cloth container and a 4" x 1 3/8" x 2 1/4" two piece plastic inner container. The outer container has a belt loop and a flap secured with two snaps. A red colored cross and 'U.S. NAVY' are stamped on the flap. The inner container was sealed with tape and has directions on the top lid and contents on the bottom lid. The aviator's kit was used with the PK-1 Pararaft Kit, and as a replacement first aid kit in the Navy M-592 Back Pad Kit.

Components:

-Compress bandage (1)
-Roll bandage (1)
-Morphine syrettes (2)
-Sulfadiazine tablets (12)
-Sea sickness pills (6)
-Triangular bandage (1) (in flap pocket of outer container)

WORLD WAR II PACKET, FIRST AID, PARACHUTE

This waterproof container had four ties to secure it to a parachute harness or other suitable, easily accessible area. It contained a Carlisle dressing, field tourniquet and one morphine syrette.

WORLD WAR II FIRST AID KITS, PERSONAL TYPE

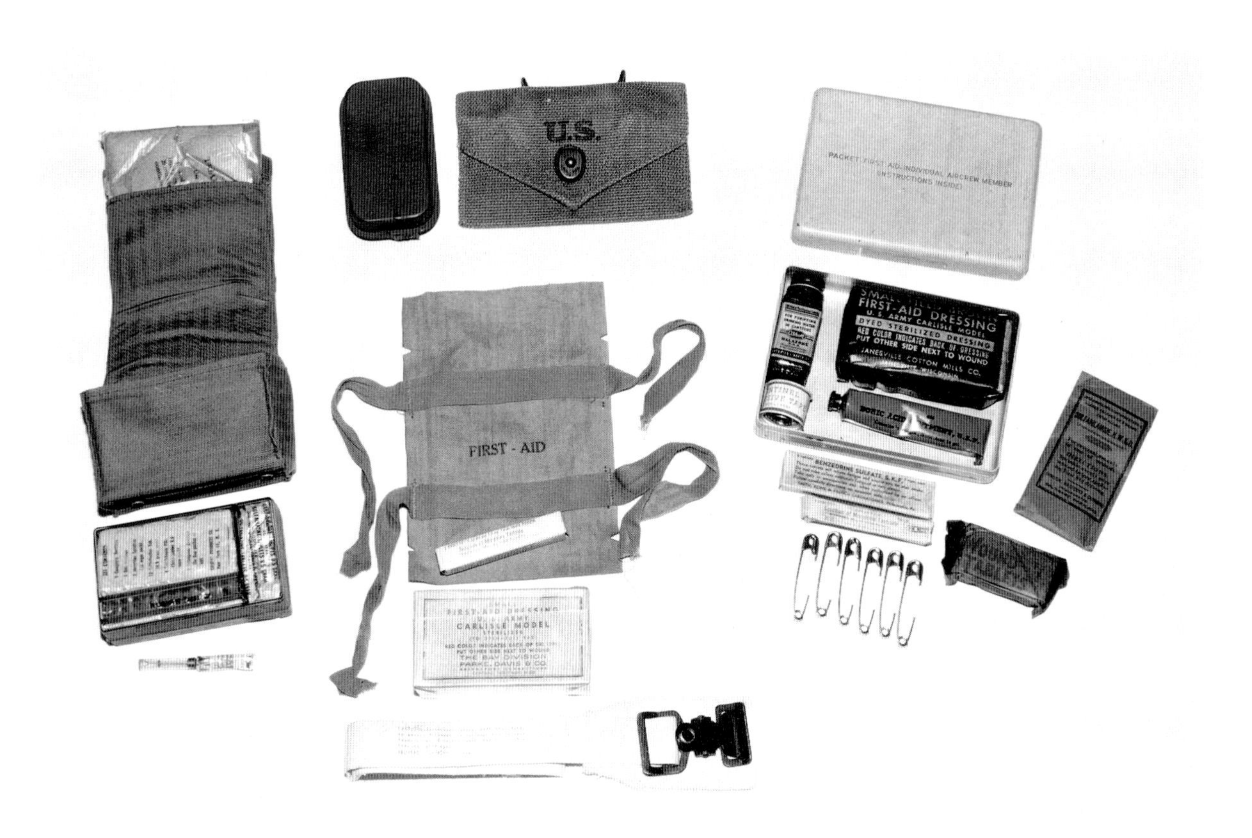

WWII individual first aid kit components (left, top to bottom): U.S.N. Aviators kit showing triangular bandage in flap, plastic case containing remaining items, morphine syrette removed from case (center, top to bottom): cloth first aid pouch and Carlisle bandage packet, parachute first aid packet with morphine syrette box, Carisle bandage and tourniquet (right): A.A.F. aluminum first aid packet showing partial components

WORLD WAR II PACKET, FIRST AID, INDIVIDUAL AIRCREW MEMBER

The Army Air Force used this 4" x 5 5/8" x 1 5/16" two-piece aluminum container and carried it in a pocket. The container was sealed with tape and has the contents and instructions printed on the inside top lid.

Components:

-U.S. Army Carlisle dressing (1)
-Morphine syrettes (2)
-Sulfadiazine tablets (2)
-Sulfanilamide powder (2 pkg.)
-Salve (1 tube)
-Benzedrine sulfate tablets (1 box)
-Water purification tablets (1 bottle)
-Safety pins (9)
-Adhesive tape (5 yds.)

STANDARD FIRST AID POUCH

A standard U.S. Army Carlisle bandage was carried in the cloth U.S. marked first aid pouch with belt hook for the pistol belt.

POST WORLD WAR II FIRST AID KITS, PERSONAL TYPE

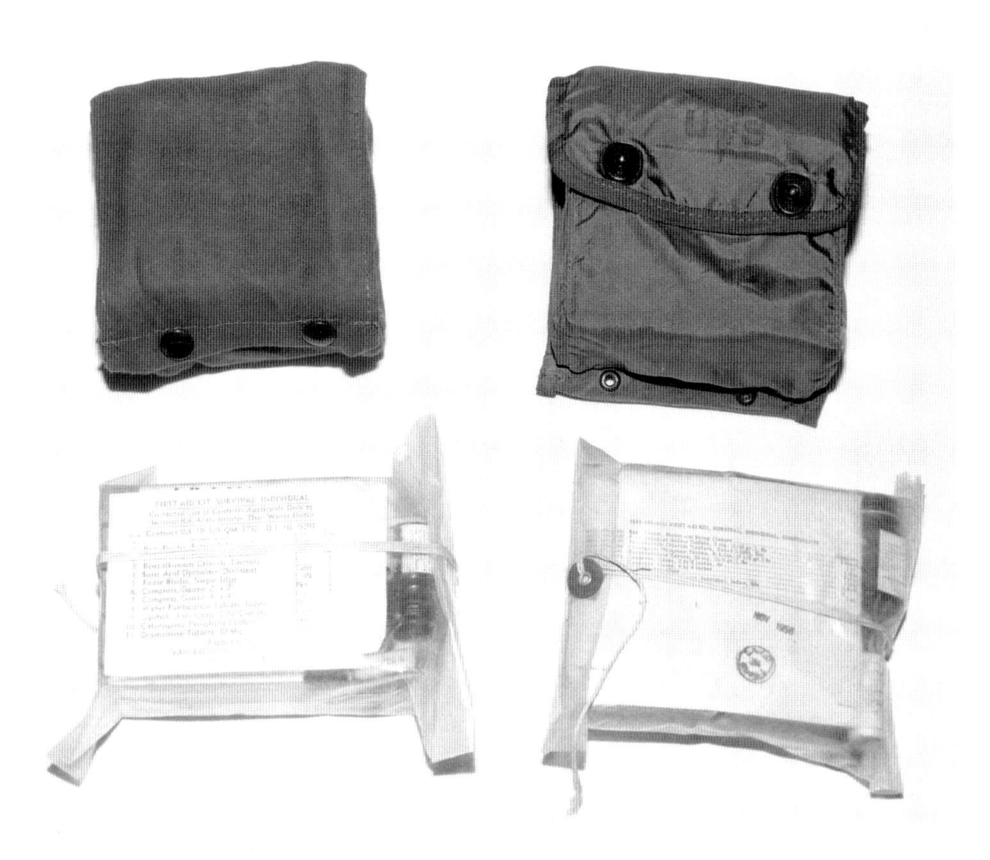

Individual seat survival kit, first aid kits (top row, left to right): Aviator's Camouflaged (1966), current issue nylon (bottom row, left to right): 1961 and 1957 dated first aid kits using vinyl bags secured with string and button closures

Left: Aviators Camouflaged first aid kit components

POST WORLD WAR II FIRST AID KITS, PERSONAL TYPE

FIRST AID KIT, AVIATOR, CAMOUFLAGED (6545-965-2394)

Components for the first aid kit:

ITEM	STOCK NUMBER
-Case, first aid kit, empty canvas, w/plastic insert (1)	6545-965-2395
-Benzalkonium chloride tincture (1 bottle)	6505-299-8183
-Oxytetracycline tablets (16)	6505-299-8276
-Meclizine hydrochloride tablets (1 package)	6505-634-7281
-Compress, gauze, camuflaged, 4" x 4" (1)	6510-200-3080
-Bandage gauze compressed, camouflaged, 3 inches x 6 yards (1)	6510-200-3185
-Bandage, muslin compressed, camouglaged, 37" x 37" x 52" (1)	6510-201-1755
-Bandage, absorbent, adhesive, 3/4" x 3" (6)	6510-597-7469
-Bottle, snap-on cap, plastic, tablet and capsule round, 4 dram (1)	6530-889-9026
-Pin, safety, curved, orthopedic, medium (4)	6530-663-1556
-First aid kit, eye dressing (1)	6545-853-6309
-Lipstick, antichap, cold climate (1)	8510-161-6205

POST WORLD WAR II FIRST AID KITS, PERSONAL TYPE

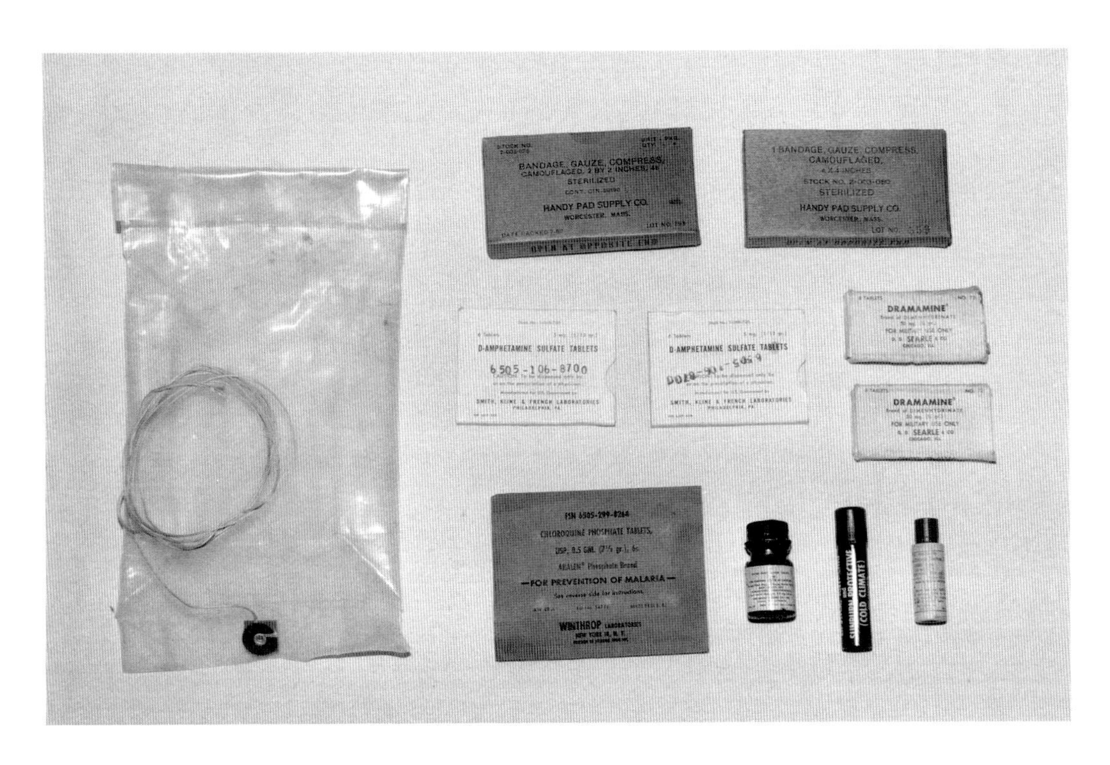

First aid kit components used in CNU-1/P container (1956)

FIRST AID KIT, SURVIVAL, INDIVIDUAL (6545-299-8313)

Components for the first aid kit:

ITEM	STOCK NUMBER
-Bag, plastic, button and string closure (1 container)	8105-299-8591
-D-amphetamine sulfate tablets, 5 mg. (2 boxes)	6505-106-8700
-Benzalkonium chloride tincture (1 bottle)	6505-299-8183
-Chloroquine phosphate tablets (1 pkg.)	6505-299-8264
-Dimenthydrinate tablets, 50 mg. (2 pkg.)	6505-261-7249
-Compress, gauze, 2" x 2" (1 pkg.)	6510-200-3075
-Compress, gauze, 4" x 4" (1)	6510-200-3080
-Lipstick, anti-chap (1)	2010-588000
-Tablet, water purification, individual, iodine, 50s (1 bottle)	8500-938000

POST WORLD WAR II FIRST AID KITS, PERSONAL TYPE

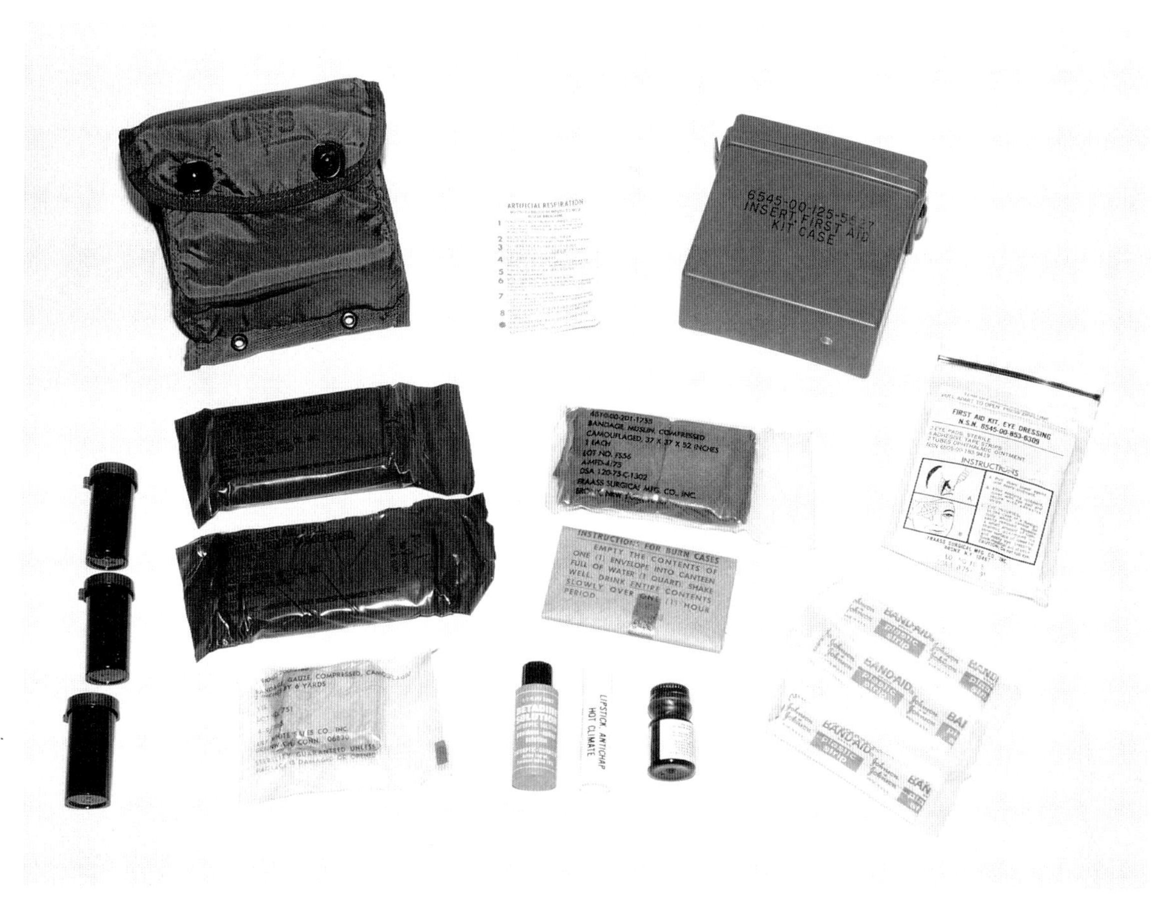

First aid kit components for the Individual Overwater, Hot and Cold Climate kits, current issue

<u>FIRST AID KIT, INDIVIDUAL</u> (6545-00-823-8165)

Components for the first aid kit:

ITEM	STOCK NUMBER
-Case, medical instrument & supply set, nylon, non-rigid, No. 8, 4 3/4" L x 2 3/8" W x 4 1/2" H (1)	6545-00-180-6239
-Insert, first aid kit case, plastic, olive drab, 4 3/4" L x 2 3/8" W x 4 1/2" H with attached cover (1)	6545-00-125-5527
-Sodium chloride, sod bicarbonate mixture, 4.5 gm, 2s (1 pkg.)	6505-00-663-2636
-Povidone-iodine solution, NF, 10%, 1/2 fl oz., (15cc), 50s (1 box)	6505-00-914-3593
-Lipstick anti-chap, hot climate, 3.7 gram, 100s (1 pkg.)	6508-00-116-1473
-Dressing, first aid, field, individual troop, camouflaged, 4" x 7" (2)	6510-00-159-4883

POST WORLD WAR II FIRST AID KITS, PERSONAL TYPE

FIRST AID KIT, INDIVIDUAL (6545-00-823-8165) continued

Components for the first aid kit:

ITEM	STOCK NUMBER
-Bandage, gauze, compressed, camouflaged, 2 inches x 6 yards (1)	6510-00-200-3180
-Bandage, gauze, compressed, camouflaged, 37" x 37" x 52" (1)	6510-00-201-1755
-Bandage, adhesive, 3/4" x 3", 300s (18)	6510-00-913-7909
-Bottle, snap-on cap, plastic, tablet & capsule, round, 5 dram (19 cc), 144s (3)	6530-00-043-1015
-First aid kit, eye dressing (1)	6545-00-853-6309
-Water purification tablets, iodine, 8 mg, 50s (1)	6850-00-985-7166

Instruction card, artificial respiration, mouth-to-mouth resuscitation

NOTE: Other items required for use with this kit may be requisitioned separately and added to FSN 6530-00-043-1015 bottles at discretion of user.

SNAKE BITE AND INSECT STING KITS

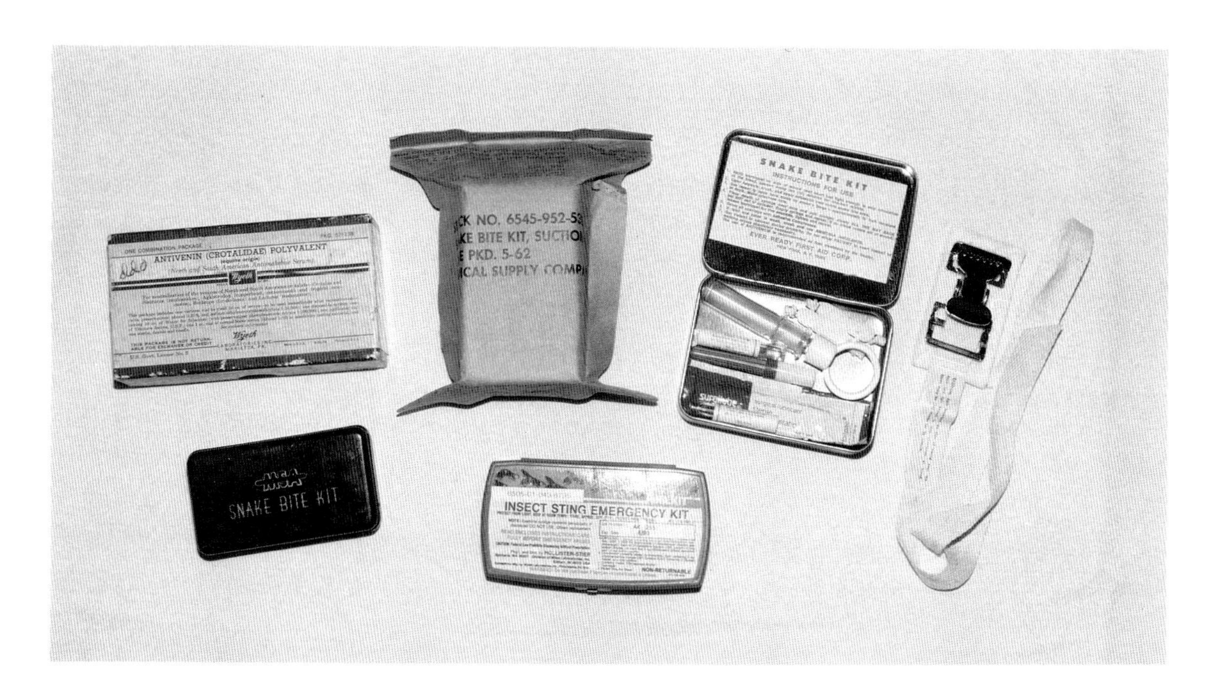

Snake bite/insect sting first aid kits (top row, left to right): 1962 dated Antivenin kit, 1962 dated snake bite kit in waterproof wrap, 1973 dated snake bite kit (bottom row, left to right): WWII MSA snake bite kit, 1990 insect sting kit

CHAPTER 4

SELECTED SURVIVAL COMPONENTS

Generally, the components used in the vests, seats and back pad survival kits can be classified into several categories: signaling devices, sustenance, protection, navigation and medical.

Signaling devices include radios, distress marker lights, mirrors, smoke/flares, pengun type flares, sea dye markers, whistles, reversible colored hats, paulins and reflective taped life raft paddles.

Sustenance items include rations, water, desalting kits, fishing kits, gill nets, firestarters and any weapons used to procure food.

Protection from the natural elements would include insect repellents and headnets, sun burn creams and shark chasers. Ponchos, paulins and parachute canopies provide shelter, and there are numerous tools in the various kits that assist in building these and other shelters.

Compasses and maps aid in navigation and the first aid kits provide for minor medical needs.

Many of thses components require further clarification due to their uniqueness, importance or just because their specific use is not readily apparent.

For example, one of the simplest signaling devices, the whistle, is included in many survival kits because its sound can carry farther than the human voice and is less tiring than constant yelling. Individuals already have less energy in survival situations because of stress and limited food intake. Any device that limits exertion, no matter how insignificant it may seem, will help prolong the individual's tolerance and chances of rescue.

WWII	VIETNAM TO CURRENT	KOREAN
-ESM-1 signal mirror -Mosquito headnet	-SDU-5/E strobe light	-Reversible hat -L-1 wrist compass -Mk-13 Mod 0 smoke/flare

SURVIVAL RADIOS

AN/URC-4 and battery

The AN/URC-4 radio set was first used in the 1950's and uses a separate battery connected by a a thirty-inch power cable. It was the recommended survival radio for use with the E-1 radio vest. It is similar to the URC-11 radio set in function. The URC-4 is a VHF and UHF transceiver operating on 121.5 and 243 Mhz and has unique dual telescoping antennas. No volume control or earphone is provided. The AN/URC-4 measures 7" x 3 1/2" x 2 1/4" and uses the same external battery as the URC-11.

SURVIVAL RADIOS

AN/URC-11 and battery

The AN/URC-11 radio set is a UHF transmitter-receiver that furnishes 2-way voice and keyed-tone communication. The complete set includes a 4" x 3" x 1 5/8" receiver-transmitter unit with separate battery measuring 6 5/16" x 3 1/4" x 1 3/8". The transmitter can operate on any preset frequency within the UHF band of 238 to 263 Mhz, although the equipment is usually tuned to 243 Mhz. The set is placed in operation by fully extending the antenna and selecting the desired type of operation by depressing push buttons on the side of the set. To discontinue operation, the antenna is returned to its case recess, allowing all control buttons to spring outward which are then locked into this poisition by a sliding catch. There is no volume control or earphone capability. The URC-11 radio set was used in the 1950's and into the 60's.

SURVIVAL RADIOS

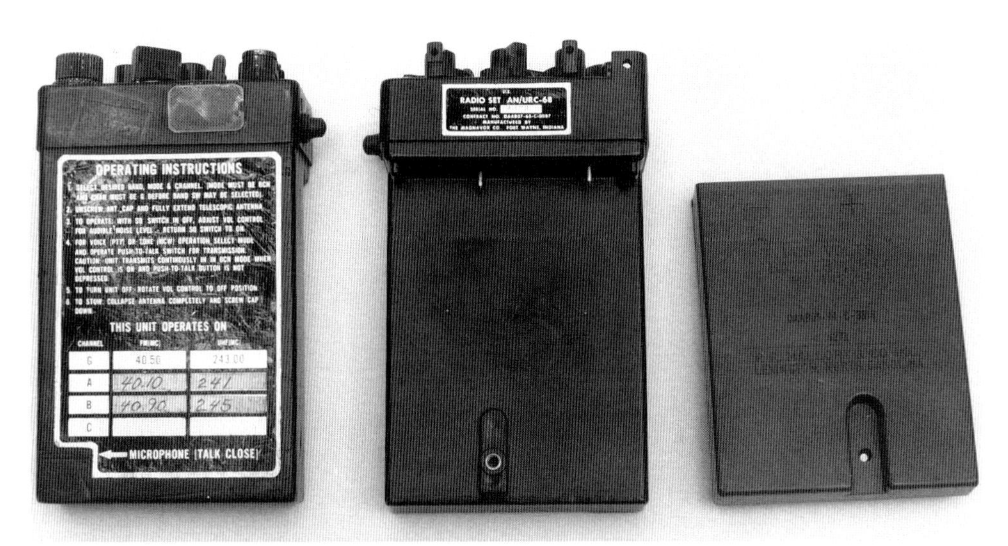

AN/URC-68 showing front, back and battery

The AN/URC-68 radio has three modes of operation: voice, MCW and beacon but has three bands: FM, UHF and SW. This 1960's vintage radio has a total of eight preset channels, four within FM (28-42 Mhz) range and four within UHF (230-250 Mhz). The 6" x 3 3/4" x 1 3/4" radio weighs 32 oz. with its large rectangular shaped, external battery with a life, depending upon the temperature, of fifteen to thirty hours in the beacon mode. Line of sight operation is approximately twenty miles at 1,000 feet altitude for homing, and ten miles in voice mode. A volume control, push to talk switch, telescoping antenna and a jack for an external antenna are also provided. An earphone is available for quiet operation.

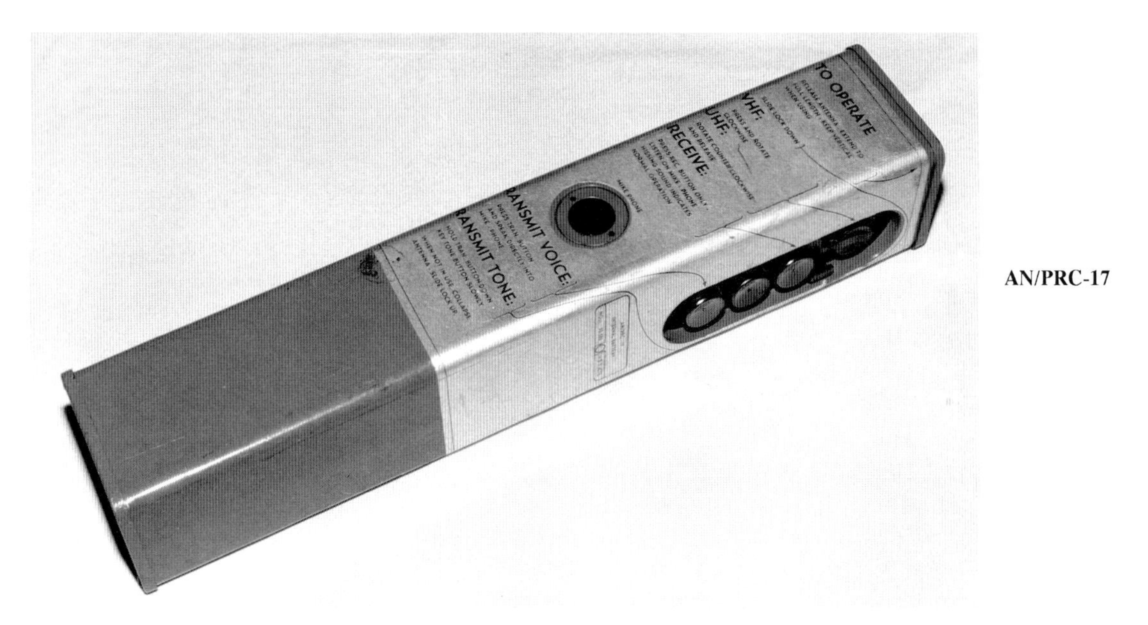

AN/PRC-17

The U.S.N. AN/PRC-17 of the 1950's was usually carried in the one-man raft kits or airdropped to individuals on the ground. Its size of 14 3/4" x 2 5/8" x 2 3/4" prohibited carrying the radio on the person while in flight. The PRC-17 furnished 2-way voice and keyed-tone communication on UHF or VHF in the 121.5 to 243.0 Mhz range. It uses a telescoping antenna and an internal battery. The range over saltwater to the aircraft at 2,000 feet altitude is approximately fifteen miles, or thirty miles at an altitude of 5,000 feet. The range over land was about half of that over water. Battery life is about twenty hours transmitting and fifteen hours receiving.

SURVIVAL RADIOS

AN/URC-64, battery and earphone case (left), ACR/RT-10 and battery (right)

The AN/URC-64 radio provides two-way ground to air communications over the entire 225-285 Mhz band. The operator can select any one of the preset channels, one of which is a 243.0 Mhz permanent guard channel. The radio measures 5 3/8" x 3 1/4" x 1 3/4", weighs 32 oz. and is covered by a watertight aluminum case. A spring loaded latch at the bottom of the case secures the internally mounted cylindrical battery. Operating instructions and Morse code are mounted on the front of the radio. A telescoping antenna, earphone, and battery test button with battery charge indicator are part of the URC-64.

The ACR/RT-10 radio is also a 1960's-dated radio and is one of the simplest to operate. This 6" x 3 1/2" x 1 1/2" radio is tuned to one frequency-usually 243.0 Mhz-although any frequency in the range of 240 to 260 Mhz can be obtained by replacement of the appropriate crystal. The RT-10 has no volume control. The on/off switch can be locked in for a tone-beacon signal for homing use during unattended periods. Line of sight operation is approximately twenty miles at 1,000 feet altitude and ten miles for voice. It has a telescoping antenna, external battery and a nylon hand strap. An earphone/acoustical coupler is available and simply covers the speaker mike for quiet operation.

SURVIVAL RADIOS

Radio beacons (left to right): AN/URT-27, AN/URT-33, AN/URT-33B

The AN/URT-21 (not shown), AN/URT-27, AN/URT-33 and 33B radio beacons are small electronic rescue units which can be carried in parachute packs, seat survival kits or the individual crewmembers. Each is a completely sealed unit with a replaceable battery and incorporate a telescoping antenna. The battery gives the beacon a 24 hour minimum operating life with transmission ranges of 40 to 60 miles obtained with aircraft at altitudes of 10,000 feet. The beacons provide a sweep-tone, modulated signal at 240.1 Mhz to enable search and rescue aircraft to home the position of the beacon. Voice transmission or reception is not possible with a beacon. When armed, these beacons activate automatically upon parachute deployment, or can be operated manually. The beacons vary slightly in size but are approximate 4" x 2 3/4" x 1 5/8" (33B).

Little information has been obtained on the RT-20A and AN/PRC-93 radios. They are both identical to the ACR/RT-10 radio except for frequencies. The AN/PRC-93 is U.S.M.C. marked and tuned to 'Code 1'. This radio has a mechanical volume control attached to the speaker; turning the speaker housing counter-clockwise restricts the opening of the mike and offers quieter operation. The RT-20A radio is tuned 251.9 MC. These radios were issued in the late 1960's.

AN/PRC-93 (left), RT-20A (right)

SURVIVAL RADIOS

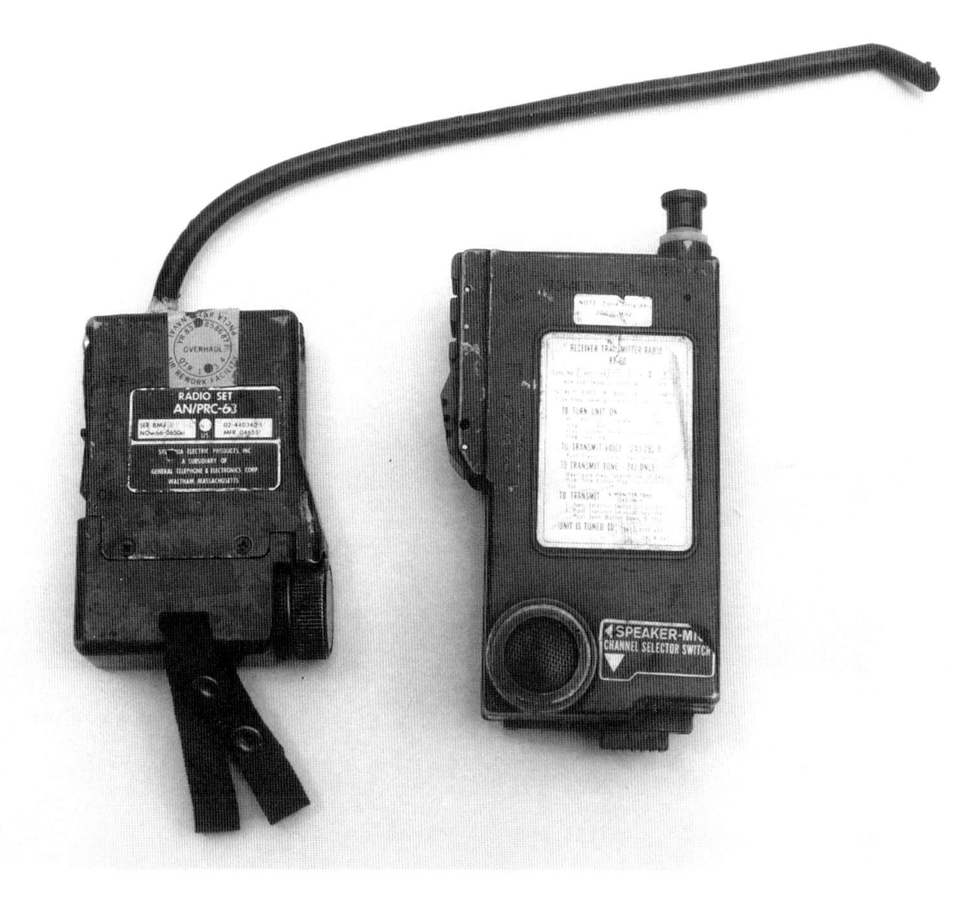

AN/PRC-63 (left), RT-60 (right)

The AN/PRC-63 U.S.N. radio measures 3 3/16" x 4 1/2" x 1 1/4" and uses the same battery as the PRC-90. It employs a flexible rubber antenna and nylon wrist strap. A sliding/toggle switch turns the radio on and off and has three functions; voice transmit, voice receive and beacon. If the voice lever is not held while turning the radio on, the beacon activates automatically. The PRC-63 operates on only one frequency, has no volume control or earphone capability, and was first used in Vietnam.

The RT-60 radio is identical to the ACR/RT-10 radio except two frequencies are employed. The frequencies 282.8 and 243.0 Mhz are selected by a sliding switch located at the bottom of the radio. Voice is transmitted on both frequencies and tone transmitted only on 243.0 Mhz. This radio was used by the Navy.

SURVIVAL RADIOS

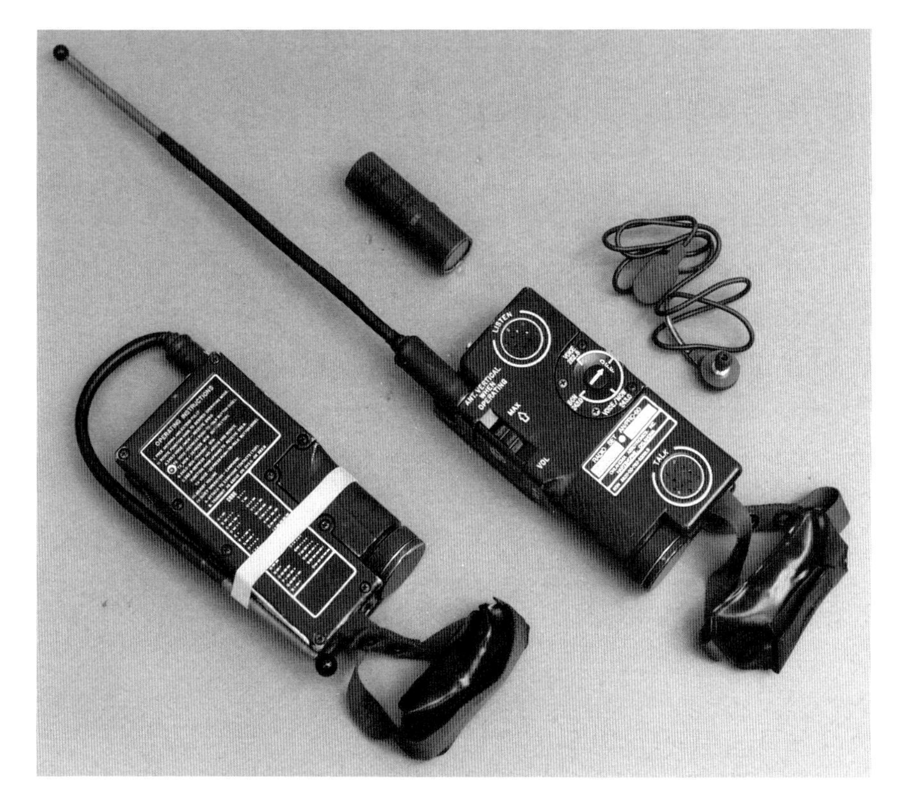

Left: AN/PRC-90 shows a battery, earphone and back of unit. A large rubber band secured the antenna for stowage in vests and kits.

Right: AN/PRC-90-1 shows a non-removable antenna, antenna securing clip. Morse code has been eliminated from the 90-1's operation.

The AN/PRC-90 radio has been in use since the late 1960's and is still in use today. Eventually, it will be superseded by the new AN/PRC-112. Its controls consist of a push to talk button, volume control, and a four position function switch to select the desired mode of operation as well as to turn the radio on and off. Voice is transmitted on either 282.8 or 243.0 Mhz, and Morse code (MCW) and the distress beacon are transmitted on 243.0 Mhz. The radio measures 6" x 3 1/16" x 1 7/16", weighs 22 oz. with a removable cylindrical battery (1" x 3") that lasts about fourteen hours in the beacon mode. Ground to air operation in the beacon mode can be 80 to 85 nautical miles at 10,000 feet altitude with voice operation up to 60 miles. Line of sight operation on the ground can be between one-half to one mile depending upon obstructions. The PRC-90 has a removable, part-flexible telescoping antenna and an earphone carried in a small vinyl case attached to the radio by a nylon wrist strap. Earlier earphone cases were constructed of hard plastic.

SURVIVAL RADIOS

AN/PRC-112

The AN/PRC-112 is the newest radio in the military inventory, replacing the PRC-90. This 28 oz., 6 1/4" X 3" X 1 1/2" radio performs as a transponder, providing personnel identification information and also allows for beacon and air to ground voice functions. It can accurately broadcast a user's location in range and bearing out to 100 nm. There are five frequencies: 243 and 282.8 in the UHF/AM band; 121.5 VHF/AM, plus any two externally programmable frequencies in the 3000 channel 255-300 UHF/AM band. The unit has a whip antenna, volume control, remote earphone and externally mounted battery with an operating life of seven hours with a 90% receive, 10% transmit ratio. A program loader is used to externally program frequencies, individual transponding codes, and mates through the battery port of the radio.

STROBES AND SIGNAL LIGHTS

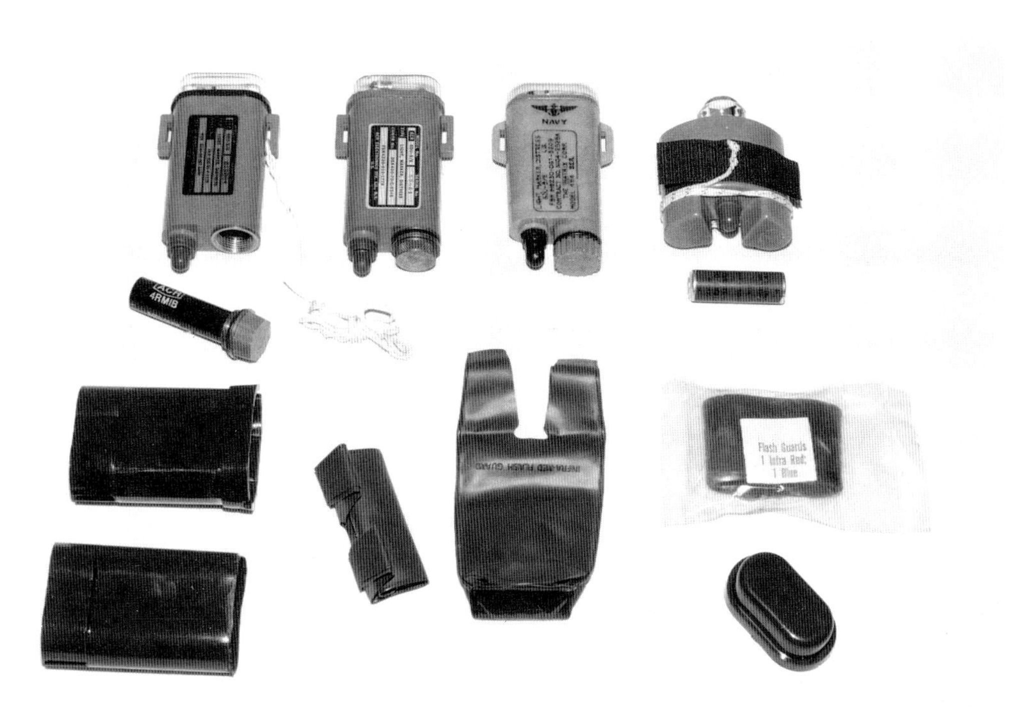

Signal lights (top row, left to right): SDU-5/E strobe dated 1989 and battery, SDU-5/E dated 1969, U.S.N. SDU-5/E dated 1967, 761-A marker light and battery with Velcro tape on light for attachment to helmet (bottom row, left to right): AGR-FG1B flashguard (1985 dated - top, 1969 dated - below), folded & unfolded vinyl red & blue flashguards contained in SRU-31/P General kit, rubber flashguard cap containing a red & blue lens also contained in SRU-31/P General kit, blue plastic flashguard cap (All flashguards are used on the SDU-5/E strobe lights.)

A variety of small marker lights have been used to help locate downed aircrews. Single 'D' cell steady burning lights were pinned to the life vests or carried in seat kits during and after World War II. Some life vests used saltwater activated lights. Steady burning lights are still available but the SDU-5/E strobe has been the most widely used since the 1960's.

The red or orange colored SDU-5/E distress marker light (strobe light), when activated, produces an intermittant flash of light visible for a minmum distance of five miles and will operate submerged in water. Several minor variations have existed since its introduction over 30 years ago. A removable flash guard (AGR-FG1B) provides directed aiming and is equipped with a blue lens so that the flashes can be distinquished from gunfire flashes. The strobe also stores in the flashguard.

The 761-A marker light is a waterproof, high-intensity steady burning light. The light has a plastic body and lens, and is powered by two mercury batteries. The Navy applies a section of Velcro tape to the SDU-5/E or 761-A for quick attachment of the marker light to the aircrewman's helmet. The helmet has a corresponding section of Velcro for a 'no hands' operation in survival situations.

Signal lights (left to right): U.S.N. WWII grey metal, Korean era grey plastic, 1970's grey plastic, 1980's orange plastic with red lens

MK-1, MK-13, MK-124
SMOKE & ILLUMINATION FLARES

Five and fifteen minute duration railroad Fusee flares (without the road spike nail) were carried in World War II seat kits and the C-1 survival vests. The 1 1/2" x 4" Mk-1 Mod 0 and Mk-1 Mod 1 smoke flares were used during World War II but the Mk-13 Mod 0 developed late in that war was the first small signaling device to combine smoke and flare.

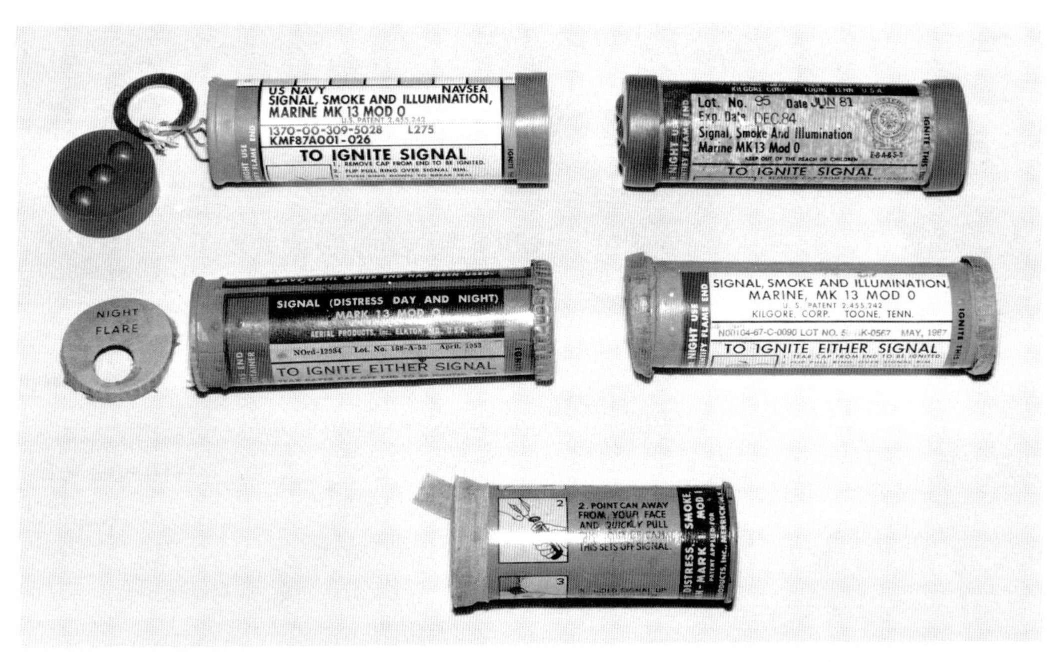

Mk-13 Mod 0 smoke/flares (top row, left to right): 1987 dated with red plastic cap removed to show pull ring (flare end), 1984 dated with plastic caps (middle row , left to right): 1953 dated with paper caps, 1967 dated with paper caps (bottom, left to right): 1945 dated Mk-1 Mod 1 smoke signal

The Mk-13 Mod 0 smoke and illumination signal is a hand actuated, 1 1/2" x 5 1/4", combination day or night distress signal. One end of the signal contains orange smoke for daytime use and the opposite end (flare) contains a pyrotechnic composition for illumination during night time use. The illumination end can be identified by a series of embossed projections located 1/4 inch below the end cap. The average burning time of the illumination signal is eighteen seconds and when used at night, it can be seen at a distance of two to three miles from an altitude as high as 3,000 feet. During daytime, the smoke end of the signal can be seen from approximately the same distance. Only one end can be used at one time and expended ends should cool before using the opposite end. A paper cap was used through the 60's to cover the ignitor pull rings on each end, and later replaced by orange and red colored plastic caps.

MK-1, MK-13, MK-124
SMOKE & ILLUMINATION FLARES

The current Mk-124 Mod 0 is similar to the Mk-13 but is intended to be operated with only one hand. The aluminum case contains four sub-assemblies: smoke candle, smoke igniter, flare candle, and flare igniter. It consists of an arming lever that must be extended to the armed position and then depressed to cock and release the firing pin.

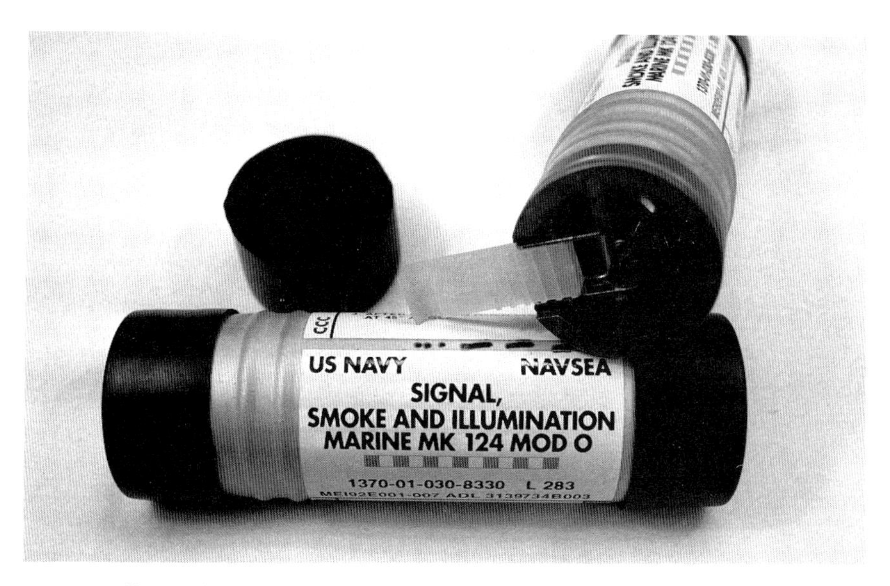

Current issue Mk-124 Mod 0 smoke/flares (top): Mk-124 with end cap removed showing arming lever extended (photo by Dave Davis)

FLARE LAUNCHERS

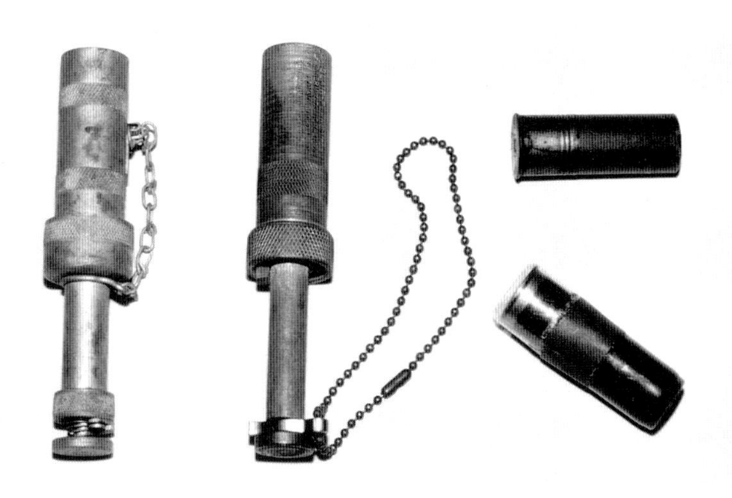

WWII era signal projectors (top, left to right): Mk-3, U.S.N. Mk-4,
10 gauge red Very shell (bottom): shell in sealed plastic container

Many types of flare launchers were employed during World War II but the Mk-3 and Mk-4 10 gauge Very projectors used by the A.A.F. and U.S. Navy during that war and into Korea were the smallest and lightest. They were also the predecessors to the current pengun and rocket type flare launchers.

They are operated by removing the assembly end of the barrel, inserting the 10 gauge Very shell, closing the safety latch, replacing the firing assembly, removing the safety, holding the projector vertically with one hand and discharging by pulling and releasing the spring loaded firing pin.

The U.S.A.F. and U.S. Army have used two distinct 'pen' type signal kits during and since the Vietnam era. The first type had a 4 inch long black or grey anodized aluminum launcher with a 2 1/4 inch long screw-in red flare. These kits were designated the M-185 or the A/P22S-1. The launcher had a knurled handled for grip and a thumb snap trigger that would fire the flares to an altitude of 200 to 250 feet for a duration of 5 to 8 seconds. The seven flares were carried in a 'tape' bandoleer and attached to an eyelet on the launcher by a lanyard. Each flare has a red plastic screw cap placed over the firing primer which had to be removed before insertion into the launcher. This signal kit was replaced in the late 60's, early 70's with the M-201 (A/P 25S-5A) 'foliage penetrator' rocket launcher signal kit. These rocket flares could reach an altitude of 1,100 feet and burn for approximately nine seconds. The flare was visible for up to three miles in the daylight and ten miles at night. The flares were contained in a red plastic retainer and press fit into the end of the launcher. The M-201 was significant because, unlike the previous type of flare, the rocket flare could better penetrate light jungle foliage.

FLARE LAUNCHERS

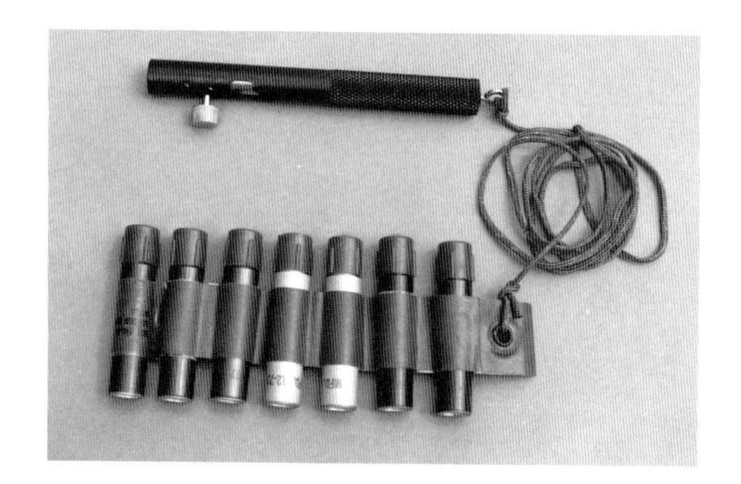

Left: 1971 dated U.S.A.F. 4 3/4" long pengun launcher with multi-colored flares - 3 red, 2 silver, 2 green

The Navy has continued to use the screw in type flares exclusively, although the rocket flare kits have been found in some Navy kits. The 4 3/4" long launcher and seven flares are similar to the Air Force flare kits except that each flare and launcher are U.S. Navy marked and designated as the Mk-79 Mod 0 signal kit. The Mk-79 is comprised of seven Mk-89 flare cartridges contained in a red plastic holder, and one Mk-31 launcher. The launcher is gray/black anodized aluminum and the flares have silver colored cases. The launcher attaches to the flare holder with a lanyard. Some of the first 'pen' style signal kits had gold colored launchers and red flares.

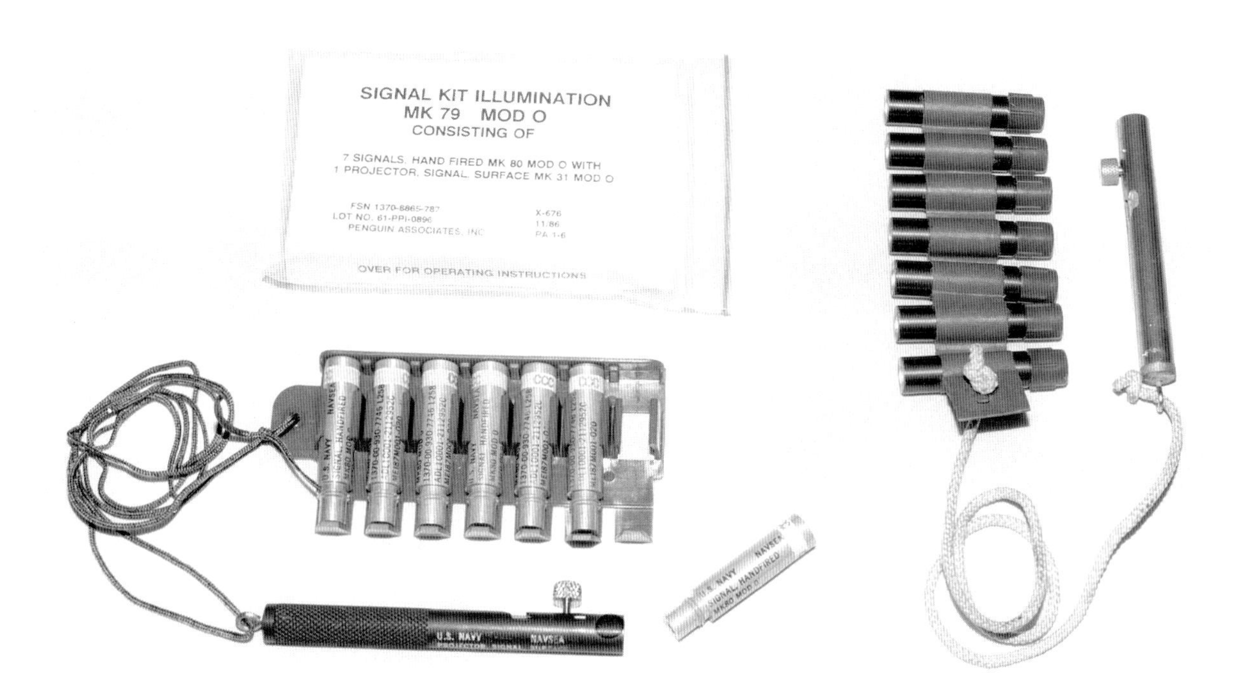

U.S.N. Mk-79 Mod 0 flares and launcher (left), early type gold colored flare launcher and flares (right)

FLARE LAUNCHERS

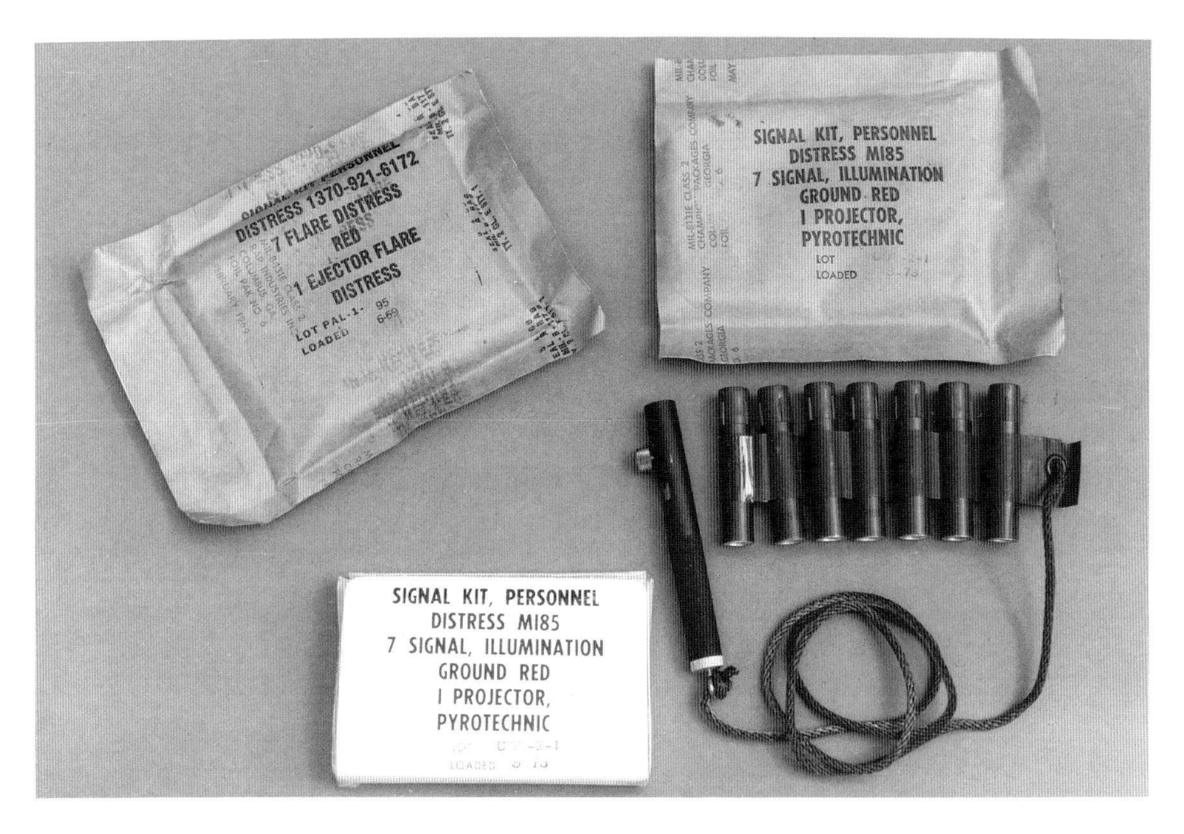

U.S.A.F. Personnel Signal kits (1969 dated on left, 1973 dated on right)

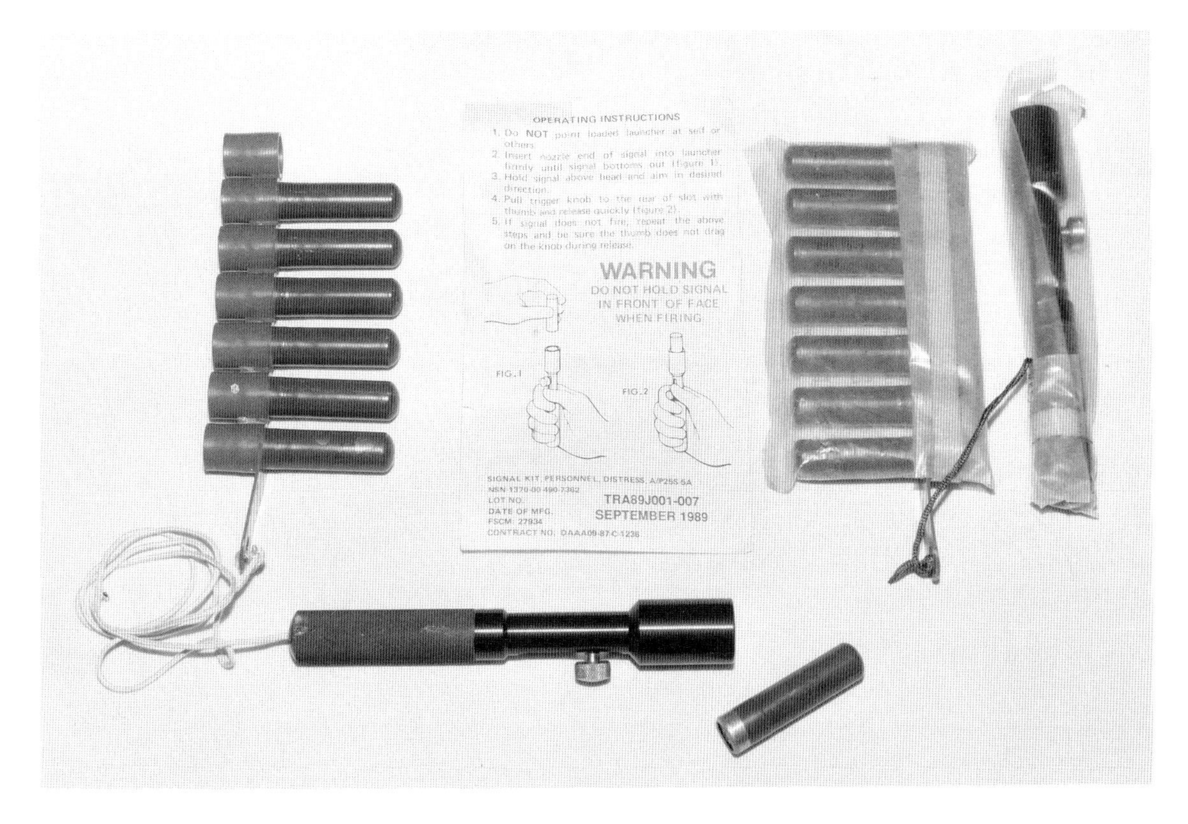

U.S.A.F./U.S. Army 'rocket' type flares and launcher (dated 1985) (right): Launcher and flares are usually wrapped in plastic and then placed in the vest or kit.

SIGNAL MIRRORS

The signaling mirror is a glass signaling instrument equipped with a retaining cord on one corner and a see through sighting hole in the center of the glass. When used in the daytime, and with good visibility, a mirror flash can be seen up to a distance of thirty miles at an altitude of 10,000 feet. Though less effective, and with shorter range, mirror flashes can also be seen on cloudy days. There have been many different styles and types of signaling mirrors manufactured as a survival item. The more common sizes are 3" x 5" and 2 " x 3" and all usually have instructions printed on the sighting side of the mirror. They have been primarily made of glass but polished metal mirrors are not uncommon.

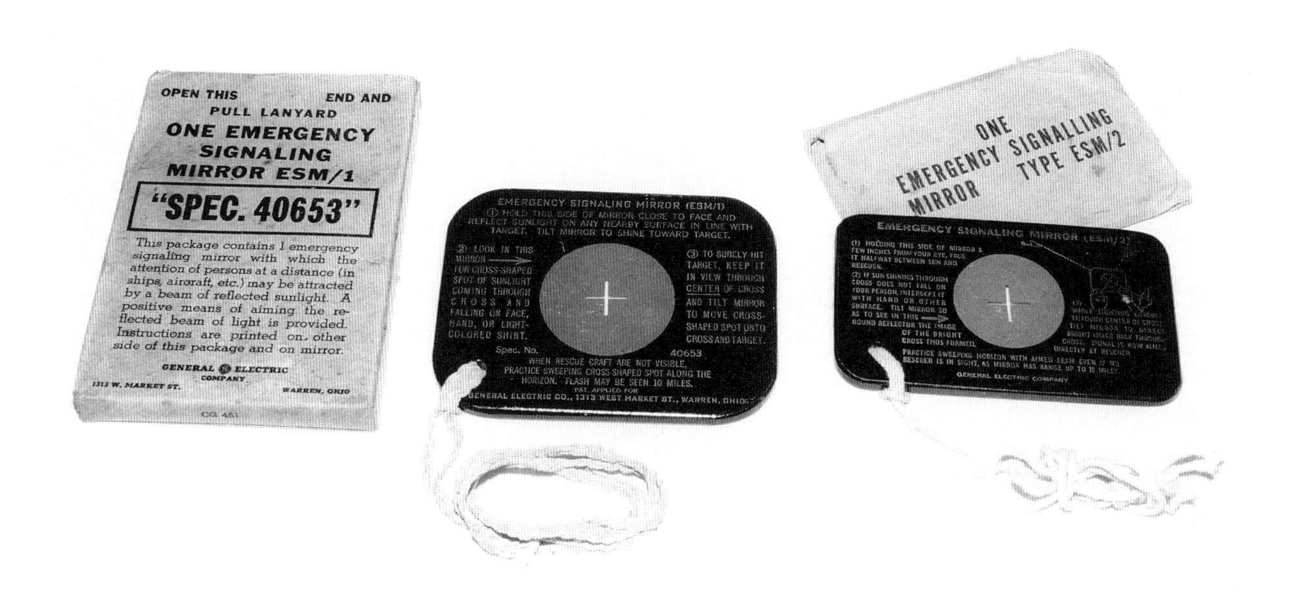

(left): 4" x 5" ESM-1 mirror used in E-17 personal kit and U.S.N. M-592 (right): 3" x 5" ESM-2 mirror used in various seat kits and C-1 vest

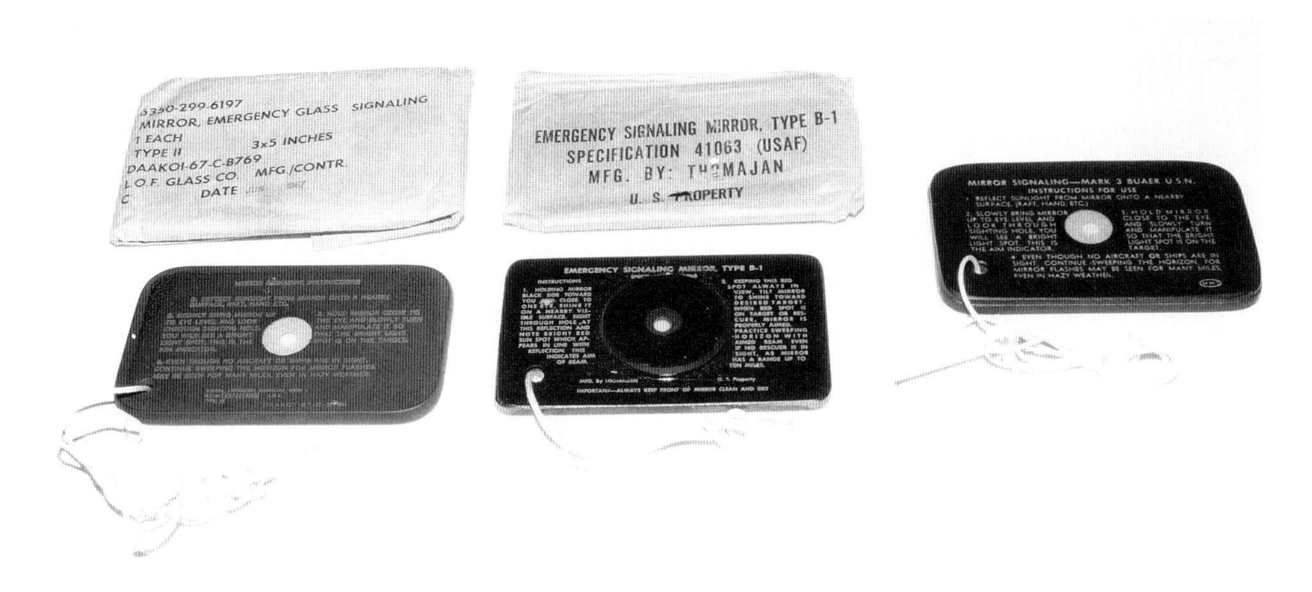

3" x 5" signal mirrors (left to right): Mk-3, Type II (1967), Type B-1 (1952), Mk-3 Buaer U.S.N. (All used in various seat kits.)

SIGNAL MIRRORS

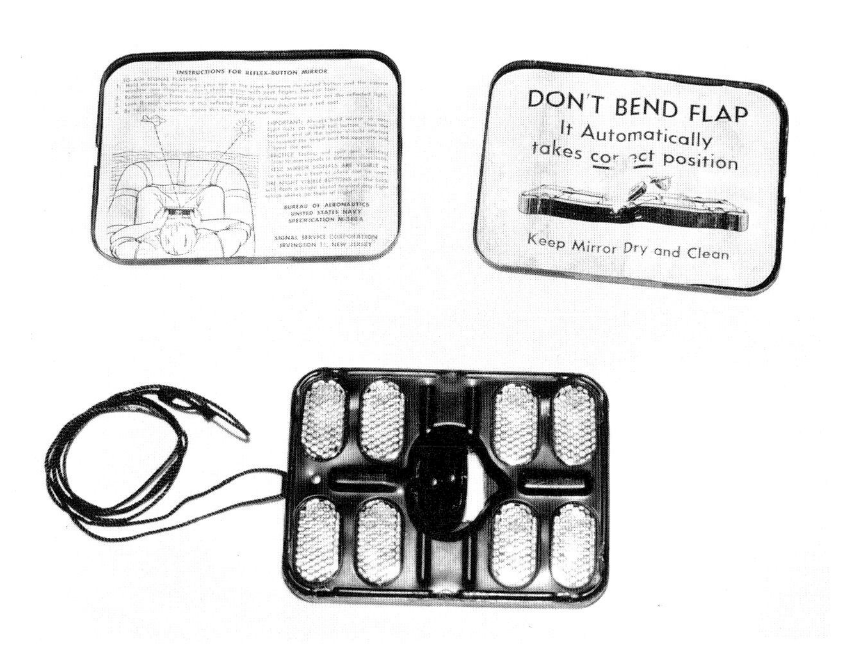

U.S.N. M-580A signal mirror used in early seat kits

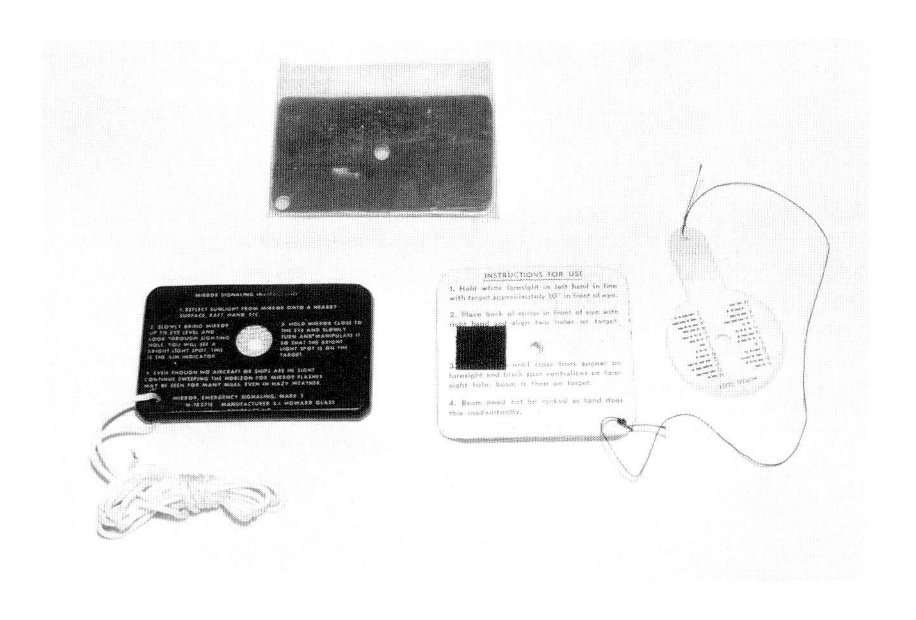

Small mirrors (top): metal 2" x 3" (bottom, left): Mk-3 Type 1, 2" x 3" used in survival vests (bottom, right): plastic type with foresight used in SRU-31/P General kits

RATIONS

WWII rations (top row, left to right): Emergency Parachute Ration (C-1 vest), Life Raft Ration, Aircrew Lunch, Hershey's Tropical bar (red label) used in E-17 type kits, 'K' rations (bottom row, left to right): Pemmican, U.S.N. Emergency Ration For Life Rafts (tablet ration), 'D' ration, water, 'K' rations

Emergency rations were produced in many different types of containers, sizes and variations in contents but were all used to provide mininum sustenance and energy for the aircrew member.

Pemmican was one of the earliest survival rations and was a mixture of crushed nuts, raisins, apples and rendered fat. The high fat content and caloric density made Pemmican an ideal survival food.

The World War II 'K' ration is the most recognizable. It came in breakfast, dinner and supper versions and each was packed in a sealed waxed inner container and outer paperboard box. Earlier outer boxes were light brown and later ones had a colored 'camouflaged' pattern. Two 'K' rations were listed in the Class 13 manual as being carried in the type E-6 Bailout Ration Bag that attached to a parachute harness. The E-7 bag was similar but contained two cans of water instead of the 'K' rations. Both of these items were part of larger survival kits. The U.S. Army Bailout Ration was specifically designed as such and a component of the E-3 Emergency Sustenance Kit. This 4" x 4" x 1 1/16" paperboard box contained chewing gum, bouillon powder, 2 oz. 'D' bars and malted milk and dextrose tablets.

Many emergency rations consist of candies, gum and vitamins for quick energy. The Aircrew Lunch, U.S.N. life raft tablet rations and standard life raft rations are among this group. Chocolate played a big part in rations and different sizes of chocolate bars are included in E-3, E-3A and E-17 personal aids kits. The U.S. Army Field Ration 'D' is a waxed paperboard box containing three 4 oz. chocolate bars and is a component of the B-1 and B-2 Army Air Force back pad kits. The Emergency Air Corp Ration is an early version of the 'D' bar and came in a metal tin.

RATIONS

The Parachute Ration and the smaller Emergency Parachute Ration contained dehydrated cheddar cheese, crackers, sugar, coffee and water purification tablets in addition to chocolate and candy. The Parachute Ration was carried in the B-4 kit and the Parachute Emergency Ration was a component of the C-1 survival vest. The metal containers of these rations were used as cooking pans or cups after the contents were removed.

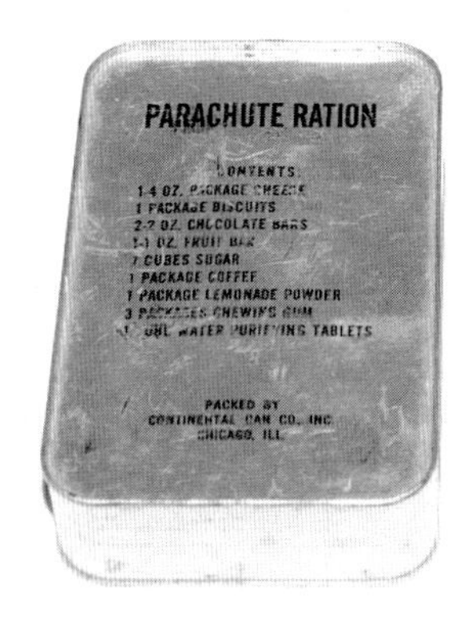

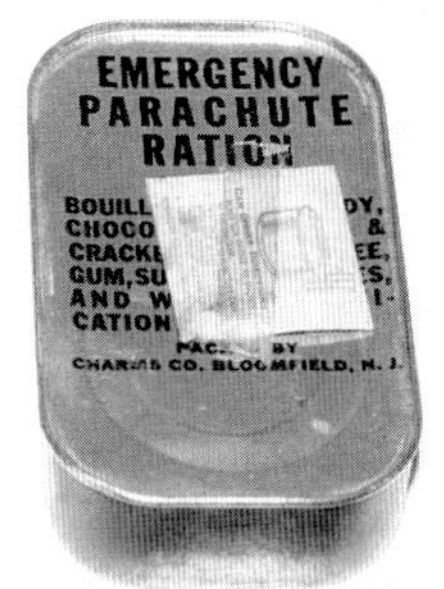

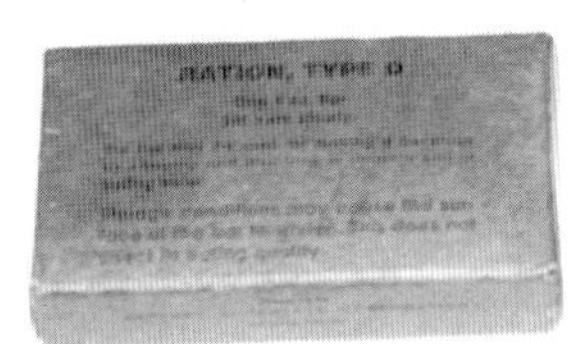
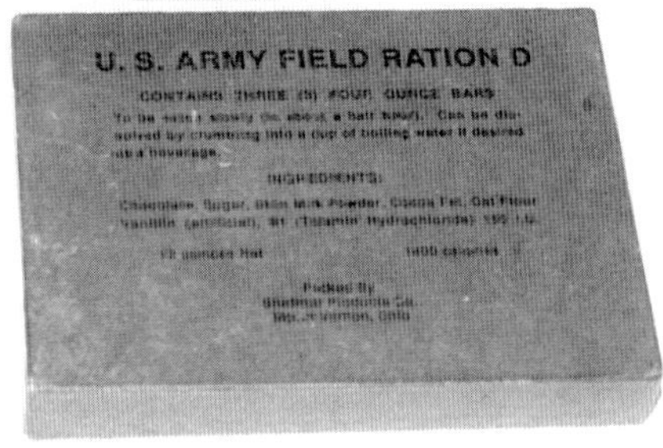

WWII rations (top row, left to right): Parachute Ration (B-4 kit) Emergency Parachute Ration (C-1 vest) (bottom row, left to right): single 'D' ration, U.S. Army Field Ration 'D' (B-2 kit)

Post World War II survival rations continued to be developed and many contained several food bars in addition to sugar, instant coffee and a soup/gravy mix. The ST food packet contained carbohydrates such as jelly bars and was designed for survival in tropical situations where stability at high temperatures was required. The SA Arctic food packet was composed of cereal bars, fruit cake bars, cheese and chocolate bars, and had a protein-carbohydrate fat ratio better suited to colder regions. The ST and SA food packets were used in the Air Force CNU-1/P container and other seat kits.

The Abandon Aircraft was a 2-part ration containing special meat food bars in addition to fruitcake bars, tea, sugar and seasonings. This packet is designed with a high caloric value for use under cold survival conditions where hard work is required and an ample supply of water is available. The Abandon Aircraft rations came in a 22 oz. and 12 oz. metal container.

The All Purpose and General Purpose (GP) food packets were designed to meet the need for a single survival ration which could be used in a number of types of survival situations. The All Purpose and GP rations replaced the ST and SA rations. These food packets came in a 'Spam' type 12 oz. can and had four concentrated food bars. The OV-1 cold climate kit contained a 'modified food packet' which was packed in a sealed plastic bag inside a paperboard box.

The first three types of GP rations came in metal cans and the current one comes in an aluminum container. As with most canned rations, opening was accomplished by using either the attached key or the standard 'P-38' opener which was taped to the can.

RATIONS

Candies and gum were carried in the many PSK's such as the individual 1 and 2-part kits, the SEEK 2, SRU-31/P, etc. and were mainly used to alleviate thirst, reduce the discomfort from hunger and to give energy.

Water was carried in seat kits in 10 oz. rectangular and cylindrical metal cans from World War II to the present and have been available for several years in 4 oz. flat foil packets that can be carried in the survival vests. Three pint and five quart water bladders are also included in vests and seat kits to carry locally obtained water. The Navy SV-2 vests uses one or two 4 oz. plastic water bottles.

Post WWII rations (top row, left to right): 2-part Abandon Aircraft Food Packets dated 1966, Artic Survival Food Packet (SA-6) dated 1957, foil wrapped drinking water, canned drinking water (middle row, left to right): All Purpose Survival Food Packet (early 60's), early General Purpose Survival Food Packet (GP) dated 1968, GP food packet dated 1983, current GP food packet in aluminum container dated 1990, current foil wrapped drinking water (flex-pak) (bottom row): components of the GP food packet: 4 different cereal bars, can opener, soup & gravy base, sugar, instant coffee, Aircraft Life Raft Food Packet (tablet ration) dated 1963

MISCELLANEOUS FOOD PROCUREMENT ITEMS

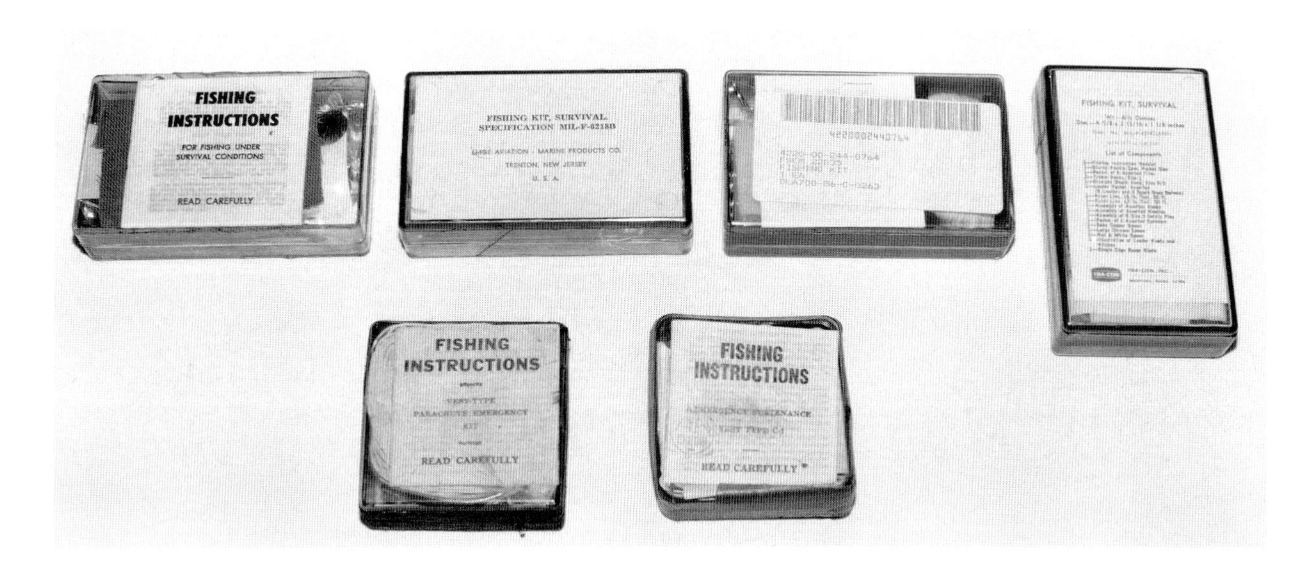

Fishing kits (top row, left to right): 1950's to early 60's kit (front and back), current issue kit (front and back) (bottom row): C-1 vest variations

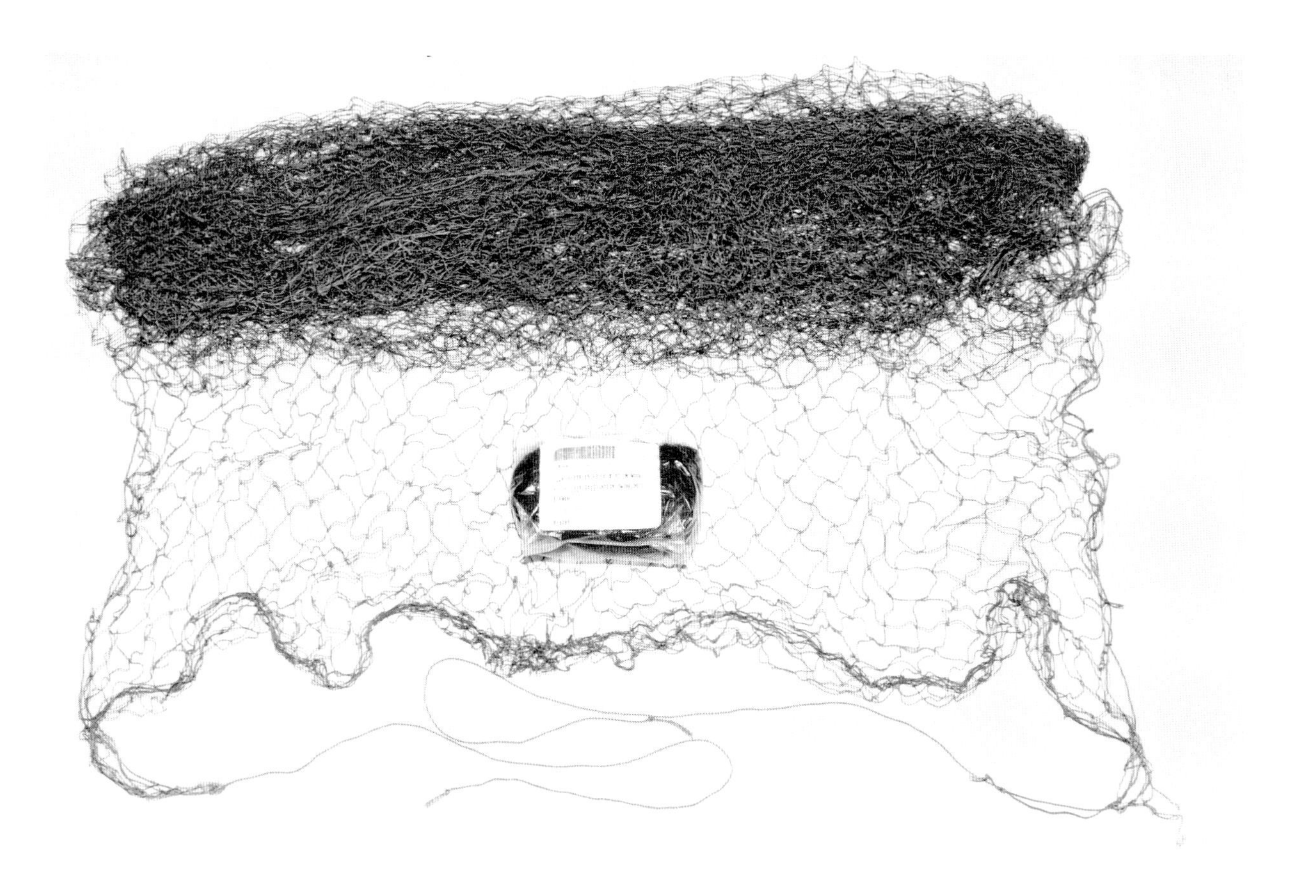

Gill net (shown wrapped in center)

DESALTING KITS

Desalting kits (top row, left to right): WWII kit dated 1945, Korean era kit dated 1952 (bottom row, left to right): Mk-2 dated 1967, current issue kit dated 1989

The seawater desalting kits carried in the various survival kits since Warld War II have provided a means of converting saltwater to suitable drinking water. Each kit consisted of several packs of chemicals, a plastic water-processing bag and tape to mend holes in the bag. World War II grey colored desalting kits carried six chemical packs. The yellow Korean war era kits had seven, and later, red and orange colored Mk-2 kits have eight packs. Each pack of chemical was capable of purifying 16 oz. (one pint) of water. The World War II, Korea, and Vietnam kits had metal containers and the current style is made of rigid plastic. A lanyard atttached to each container secured the container to the raft or individual.

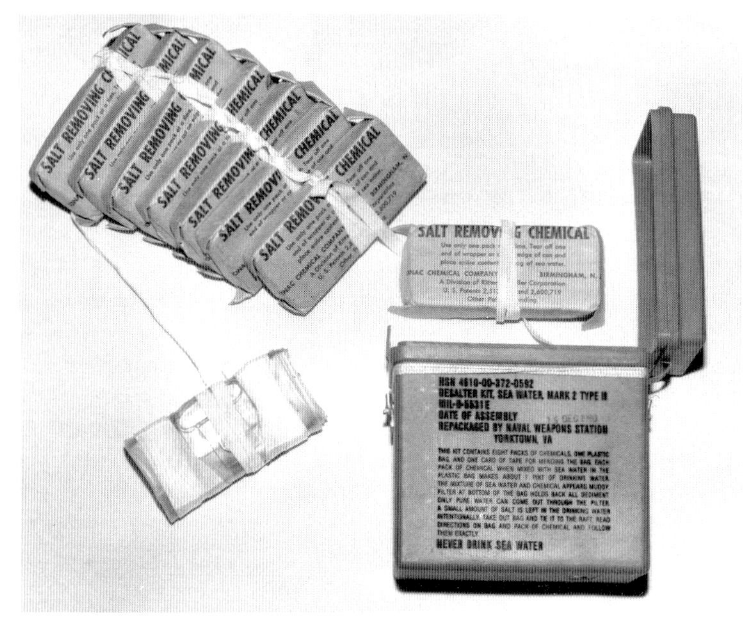

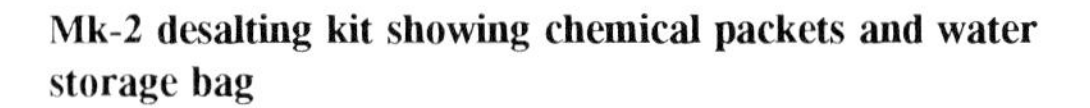

Mk-2 desalting kit showing chemical packets and water storage bag

FIRE STARTERS AND MATCH CASES

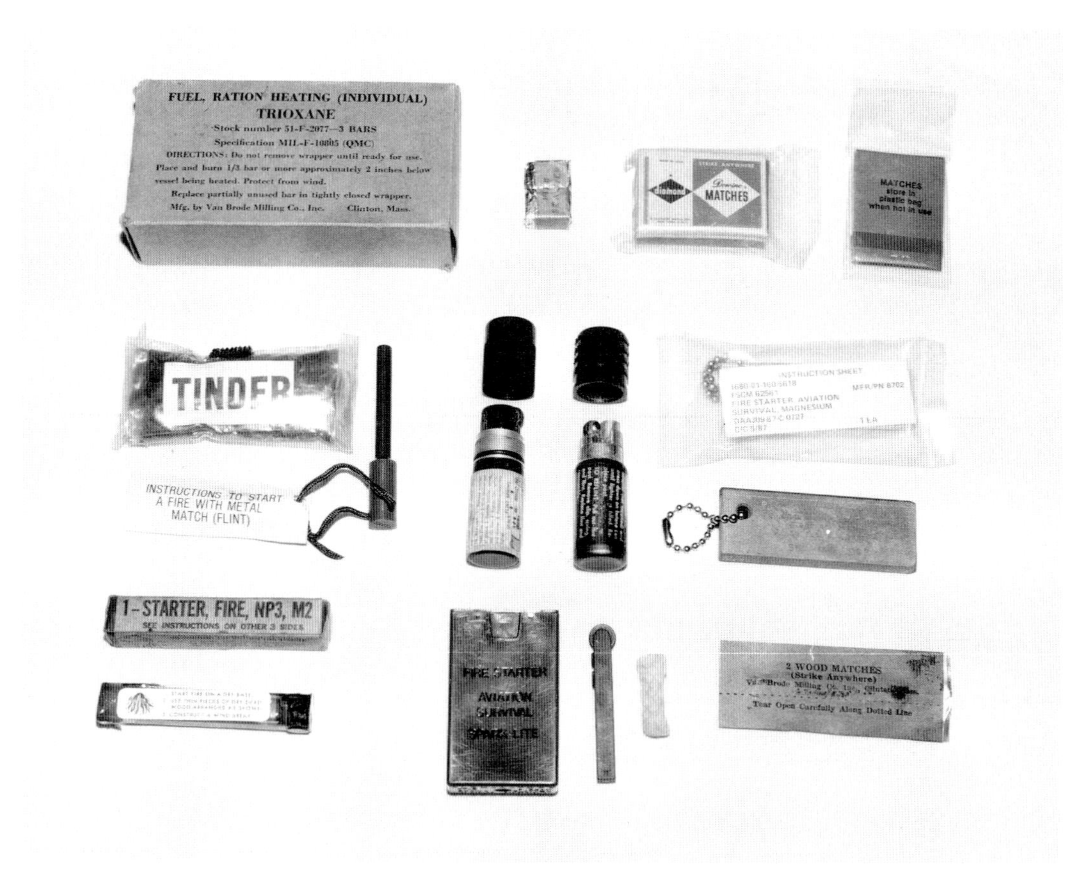

Fire starters (top row, left to right): trioxane fuel, fire starter from SRU-16/P, matches in waterproof wrap (various seat kits), matchbook in waterproof wrap (various individual personal kits) (middle row, left to right): tinder and metal match from SRU-31/P, butane lighter from OV-1 and SRU-21/P vests, butane lighter from Hot-Wet 2-part kit, magnesium fire starter wrapped and unwrapped (bottom row, left to right): M-2 fire starter and box, 'Spark Lite' container showing sparking flint and tinder, wood matches in waterproof wrap

Many types of fire starters have been issued for survival use but the match case containing wooden matches remains the simplest and most popular. Though other types of fire starters are part of survival kits, this item is virtually always included. Variations exist in match cases as well. Colors can range from the brown and red World War II models to different shades of green.

Most cases have a sparking flint on the bottom, a screw top that seals against moisture, and early versions have a small compass in the cap. Matches can be found in other types of containers such as the common matchbook and box and in small paper or plastic packets. All of these containers are sealed against moisture. In addition to the match case, alternatives are often included. The magnesium fire starter is a very effective tool. It does not need to be kept dry, nor is it fragile, and when used sparingly, it can provide many hundreds of successful fire starts. The metal match and tinder is another backup to matches, and newest item, the Aviator's Fire Starter, is growing in favor. Each of these items provide their own tinder to aid in fire starting and use a flint sparking strip or wheel to ignite the tinder.

Yet another backup is the butane lighter which is specially designed to be carried at altitude and functions as a normal lighter once the screw cap has been removed. Lastly, the M-2 fire starter is a small plastic container (3" x 1/2" x 1/2") of NP3 'napalm' with its own ignitor. It is useful in starting fires in humid and damp areas.

FIRE STARTERS AND MATCH CASES

Trioxane fuel tablets have been included in many seat kits and can be used as tinder for fires or on their own to heat liquids. These fuel tablets are primarily composed of metaformaldehyde which is toxic and should not be ingested, nor the fumes inhaled.

Match cases (top row, left to right): 1960's green case, 1980's o.d. case, current container as used in Individual Cold and Hot Climate kits (bottom row, left to right): WWII metal Marples type, clear plastic, standard type, combination compass and match case for C-1 vest, match case with compass in lid, U.S.N. match case with compass in lid and lanyard

SURVIVAL RIFLES

The M-1 and M-1A1 carbine and some commercial type weapons, such as the plastic stocked Stevens .22/.410 over and under rifle were used in the larger survival kits of World War II. The MA-1, M-4 and M-6 rifles were specifically designed in the late 40's as survival weapons and can be folded or dismantled to be included in the one-man seat kits. The MA-1, also known as the AR-5, is a bolt action .22 Hornet rifle similar in design to the current AR-7 civilian model. Based on available information, the MA-1 rifle was short lived. The M-4 and M-6 rifles, however, were used through the 1960's and 1970's.

M-4 (T-38) RIFLE

Left: M-4 survival rifle

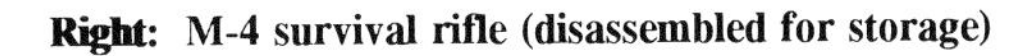

Right: M-4 survival rifle (disassembled for storage)

The M-4 is a .22 caliber hornet, bolt action rifle with a detachable five-round magazine and an adjustable rear sight. The rear sight is adjustable for both windage and elevation. Earlier models had a leaf sight located near the rear of the barrel. The model designation is stamped on the left side of the receiver, the serial number is stamped on both the detachable barrel and the receiver.

The M-4 survival rifle may be dismantled for storage in various survival kits. The barrel and bolt can be removed and the shoulder brace (stock) collapsed into the receiver. The rifle weighs 4 pounds, has an overall length of 32 1/2" with the shoulder brace extended, a barrel length of 14 inches and an effective range of 100 yards.

SURVIVAL RIFLES

M-6 (T-39) RIFLE

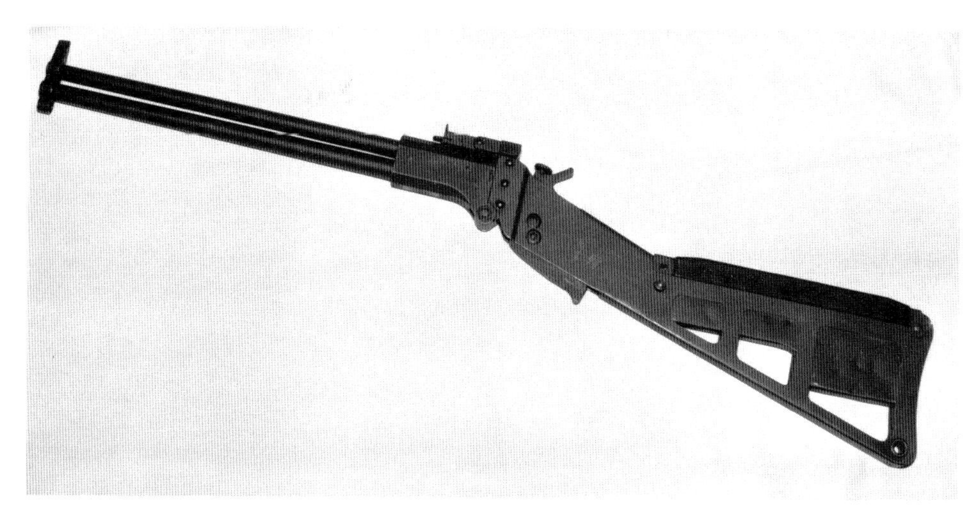

M-6 survival rifle

Right: M-6 survival rifle (folded/stored position)

The M-6 is a single shot shoulder-fired weapon consisting of an over and under combination .22 caliber Hornet rifle on top and a .410 gauge shotgun below. Nine .22 rounds and four .410 shells can be stored in the cartridge box assembly in the sheet metal stock. The front sight is a preadjusted blade type sight. The 'L' shaped rear sight, also preadjusted, consists of two legs, a peep sight for the .22 and an open sight for the .410, each marked for the specific caliber or gauge. A selector is located in front of the hammer for firing either barrel. A vinyl plastic molded covering on the stock protects the operator from the cold metal surface when firing. The model designation is stamped on the left side of the receiver and the manufacturer's name is on the right side. The M-6 weapon is 28 1/4" long and 15" in the folded/stored position. It weighs 3 lbs. 12 oz. and has an effective range of 25 yards for the .410 and 50 yards for the .22 rifle.

SURVIVAL RIFLES

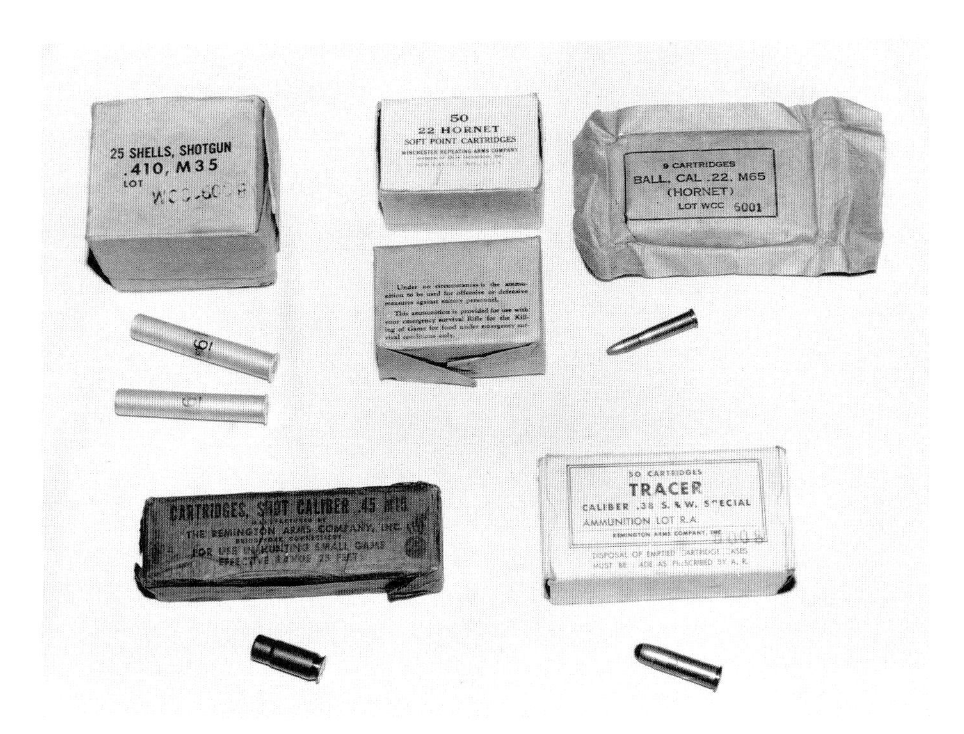

Survival ammunition (top row, left to right): .410 M35 shotshell, .22 caliber Hornet (top box out of waterproof wrap - 50 rounds), .22 caliber Hornet in 9 round package (bottom row, left to right): .45 caliber shot ammo, tracer ammo

The .410 gauge round contains #6 shot in a 2 7/8" long aluminum case. The .22 caliber Hornet uses a soft point bullet. Hollow-point bullets were not authorized to be used.

Note below the label on the 50 round boxes of .22 caliber Hornet ammunition:

"Under no circumstances is the ammunition to be used for offensive or defensive measures against enemy personnel. This ammunition is provided for use with your emergency survival rifle for the killing of game for food under emergency survival conditions only."

SURVIVAL TOOLS AND KNIVES

A-2 tool showing shovel blade and handle attached to saw/knife blade (left): SRU-18/P tool showing canvas holder, spade, ax/hammer/saw, handle

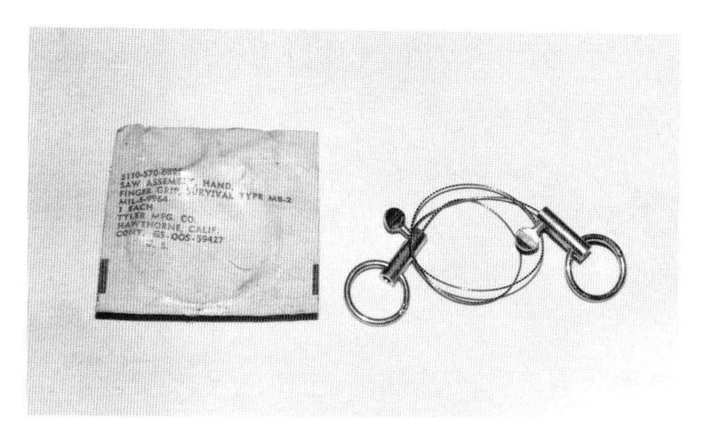

MB-2 finger saw

SURVIVAL TOOLS AND KNIVES

Left: Type IV Survival Tool showing case, manual, instructions, sharpening stone and case, burning lens, 'machete'

Bottom: Machetes (left to right): 10 inch folding blade used in B-2, and B-4 kits; 10 inch nonfolding blade (wooden handle) used in U.S.N. M-592 kit; Type A-1, 10 inch folding blade with leather sheath and sharpening stone

SURVIVAL TOOLS AND KNIVES

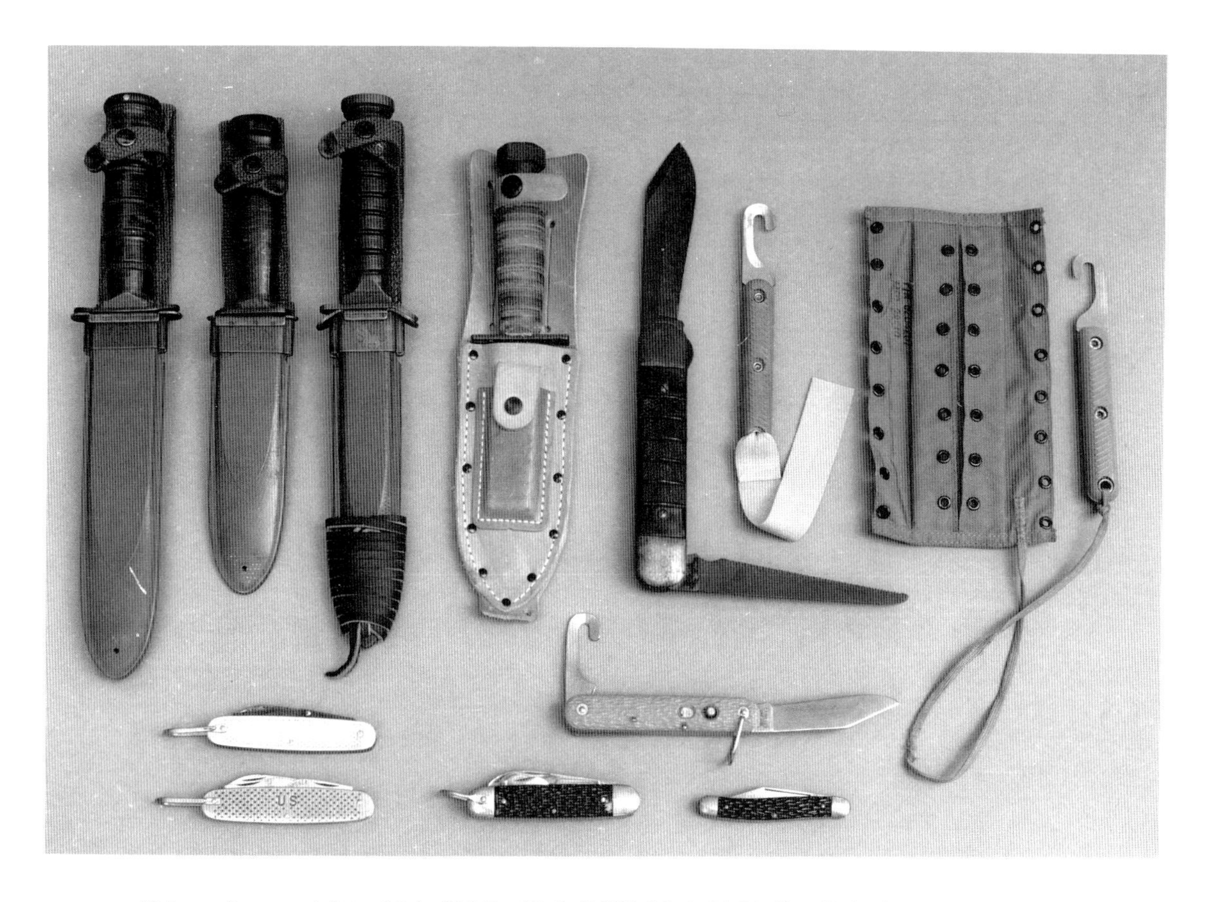

Knives (top row, left to right): U.S.N. Mk-2, U.S.N. Mk-1, M-3 knife, 5" pilot's knife and sheath, C-1 vest knife/saw (A similar knife was used by the U.S.N. and had an attached lanyard 'D' ring.), shroud knife (used in vest), shroud knife and container attached to parachute riser. (bottom row, left to right): metal sided pocket knives (WWII on top, 1967 dated on bottom), 'boy scout' knife from C-1 vest, MC-1 orange handled shroud/switch blade knife, small pocket knife from SRU-16/P

COMPASSES

WWII compasses (top to bottom): wrist type (liquid filled), U.S. Army Engineers (left): watch compass found in B-2 and B-4 kits (right): minature compass found in E-3 and E-17 kits, C-1 vest compass/match case (left): U.S.N. compass/match case

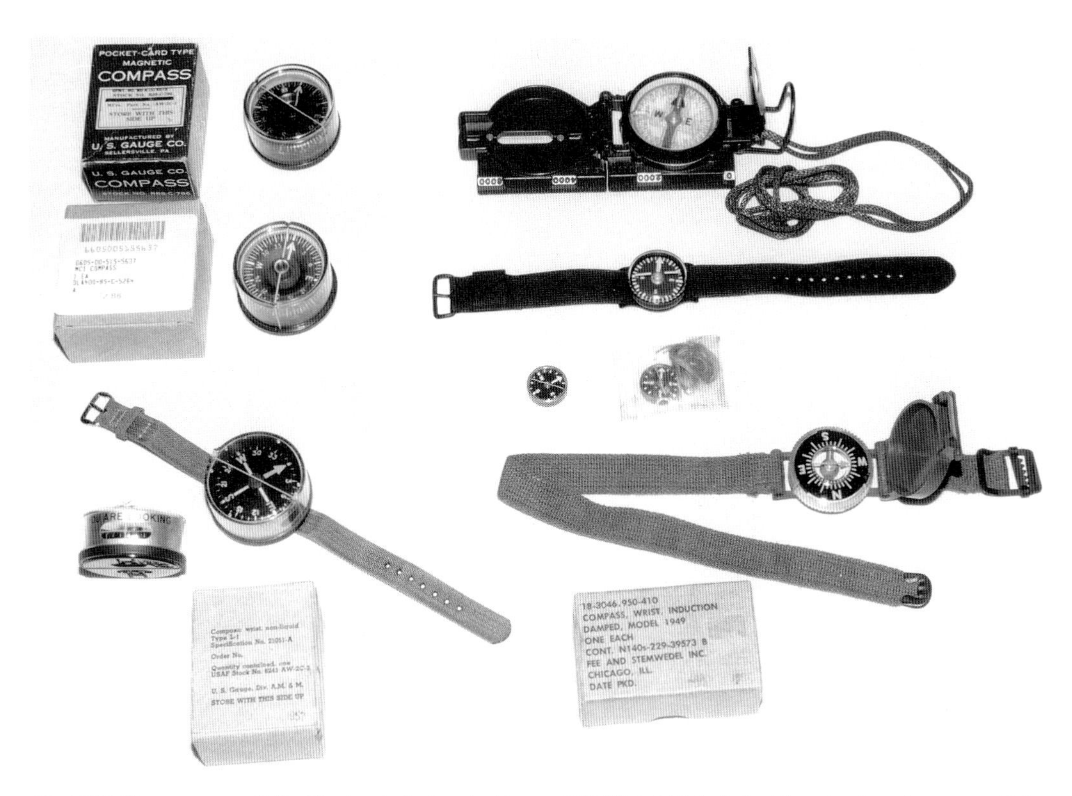

Post WWII compasses (left side, top to bottom): Bauer AW-2C-1, MC-1, L-1 with and w/o wrist strap (right side, top to bottom): standard lensatic, wrist compass found in SRU-31/P General kit, button compass, button compass with lanyard in wrap found in TAC kits, M1949 wrist compass

SURVIVAL MANUALS

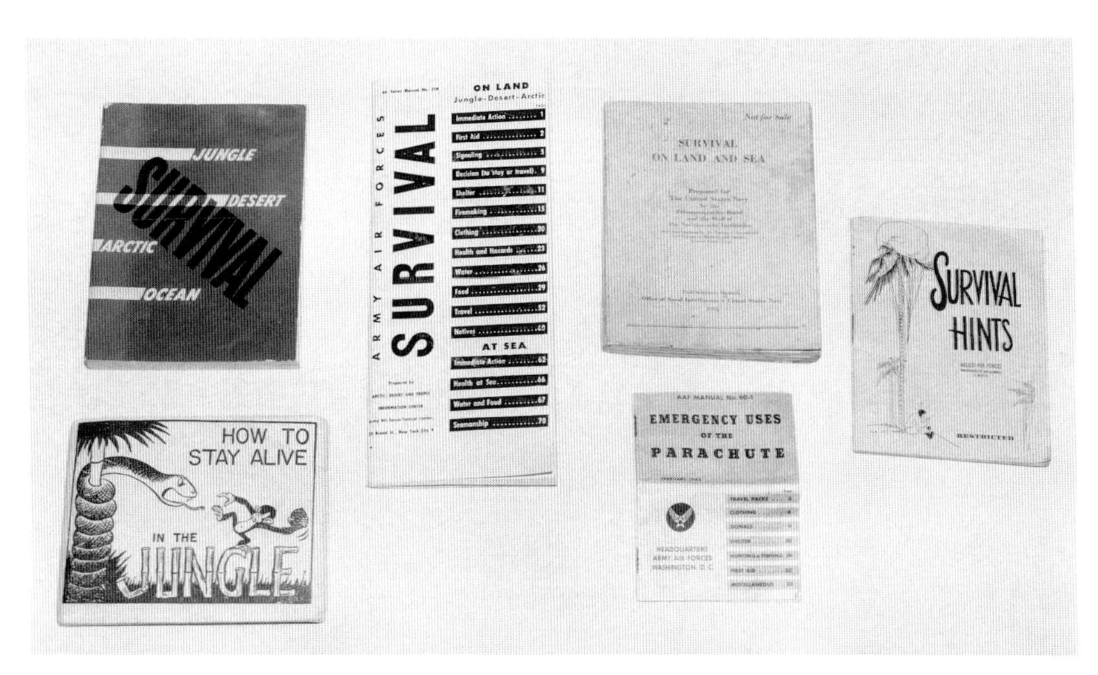

WWII survival manuals (above and right)

Post WWII survival manuals (below)

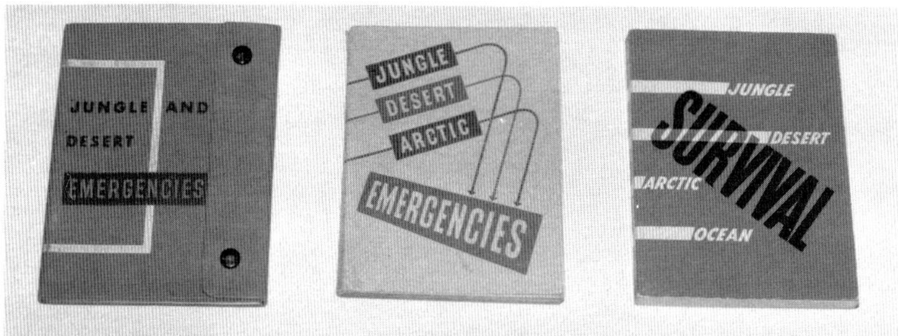

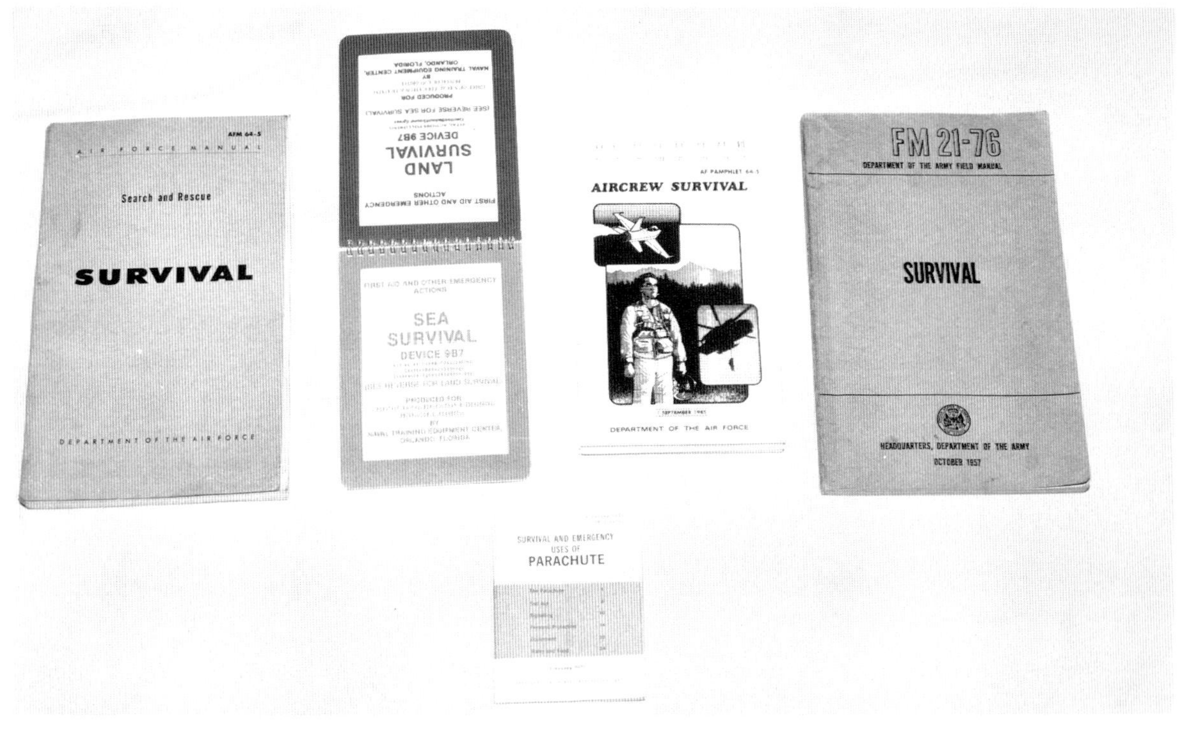

PERSONNEL LOWERING DEVICES (PLD'S)

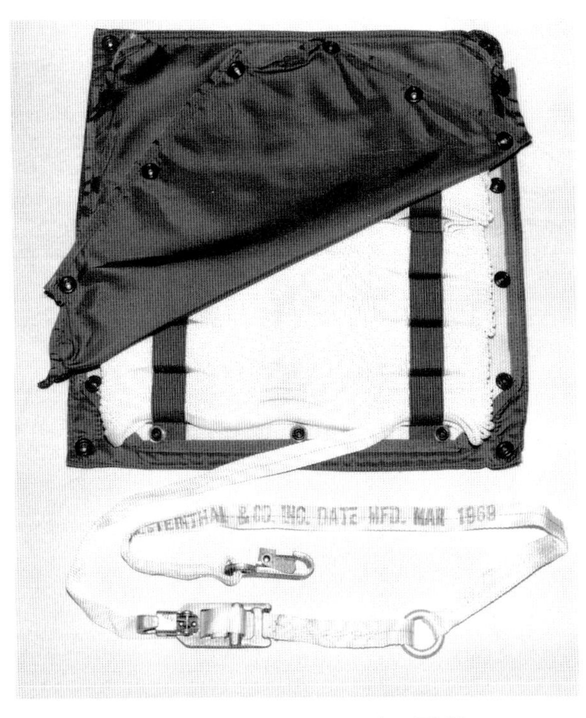

Personnel Lowering Device (PLD)

LABEL

PERSONNEL LOWERING DEVICE
TYPE PCU 10/P FOR BACK PARACHUTE
USAF PART NO. 66F 1701
ORDER NO. F33657-67-C-0946
M. STEINTHAL & CO.,INC.,N.Y.C.
U.S. PROPERTY
DATE OF MFR.
FSN 1670-897-8730

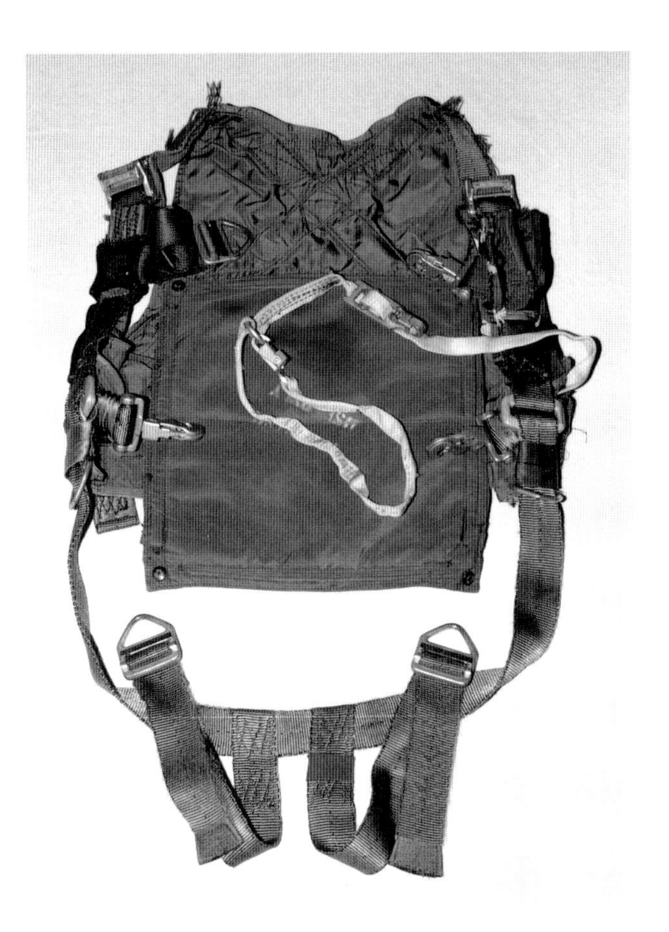

PLD shown attached to parachute torso harness

The Personnel Lowering Device (PLD) allows an aircrew member to lower himself to the ground in the event of landing in trees, power lines or rough terrain. The PLD is designed to be an integral part of the parachute harness and replaces the standard back pad when installed. The PLD line is 2,300 pound tensile strength tubular nylon, 150 feet long. Older PLD's are natural in color except for the last 25 feet, which is dyed yellow to indicate the approach of the end of the line. Newer PLD's are dyed sage green and the last 25 feet have black stripes. The PLD hardware is constructed of cadmium plated, high grade steel. The line is fed through the breaking device, allowing the aircrew member to reach the ground safely.

Several variations exist, some of the nomenclature being the PCU-11/P, PCU-9/P, PCU-21/P and the PCU-10/P. Each type of PLD is applicable to a different type of parachute harness.

For example, the PCU-21/P incorporated a lumbar pad and was sewn onto the parachute torso harness. This PLD was used with F-4 Phantom II aircraft, replacing the lumbar back pad on the Martin-Baker ejection seat.

ONE-MAN LIFE RAFTS

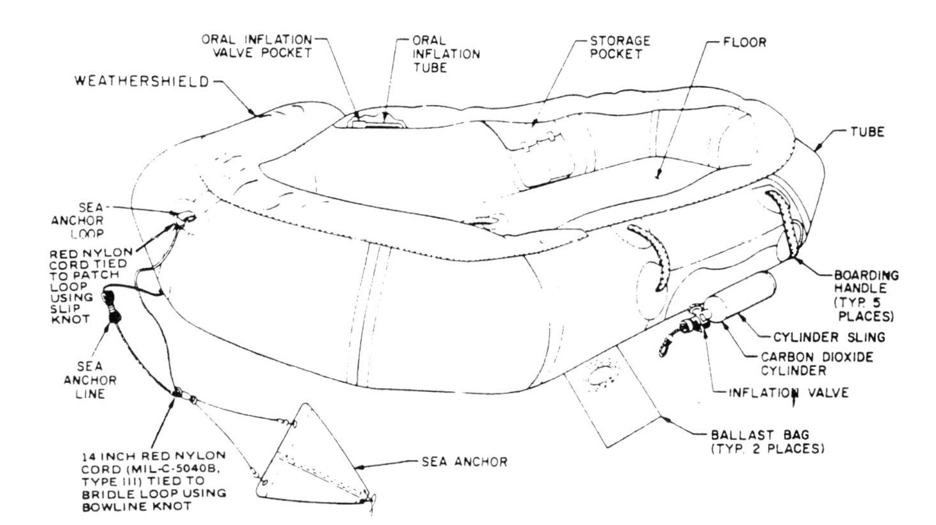

Left: LRU-3/P life raft

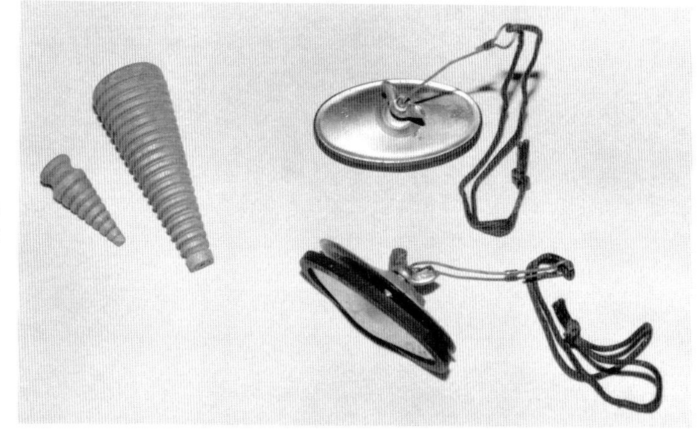

Right: Life raft repair plugs (left): early wooden 'bullet hole' type showing size variations, (right): metal and rubber plugs

One-man life rafts have been produced in various styles, materials and colors since World War II. Each raft shares a single compartment floatation tube type construction inflated by a CO2 cylinder; a sea anchor; and usually storage pockets for emergency equipment. Emergency equipment could include a flexible bailing bucket, sponge, repair kit or repair plugs, paulin, sail, paddles, sea dye marker or shark repellent. Later type wooden paddles had reflective tape on one side as a signaling aid. The sea dye markers dispensed a yellow-green color that would last about one hour in calm seas and could be seen at a distance of three to five miles at an altitude of 3,000 feet in clear weather.

The rafts were inflated manually by pulling the inflation assembly actuating lanyard and manually opening a CO2 valve or automatically by gravity drop on survival kit actuation. After inflation, the raft could be topped off through the oral inflation tube/valve. Handles were provided at various points on the raft to aid in boarding and to secure survival kits. Most one-man rafts resemble the LRU-3/P depicted above. Some early models did not have an attached spray shield as this was part of the raft kit and had to be attached after boarding. The LRU-6/P raft is different from the standard types in that the floor and canopy could be inflated to provide insulation against the cold.

The A.A.F. and Navy used the AN-R-2 and C-2 series of one-man life rafts during and after World War II. The Air Force continued with life rafts designated MB-4, LR-1, LRU-3/P, LRU-4/P, LRU-6/P and is currently using the LRU-16/P. The Navy also used the PK-1 during World War II, progressed to the PK-2 and is currently using the LR-1 one-man raft. The Army has used rafts of both Air Force and Navy designation.

BLOOD CHITS, ESCAPE MAPS, PHRASE BOOKS

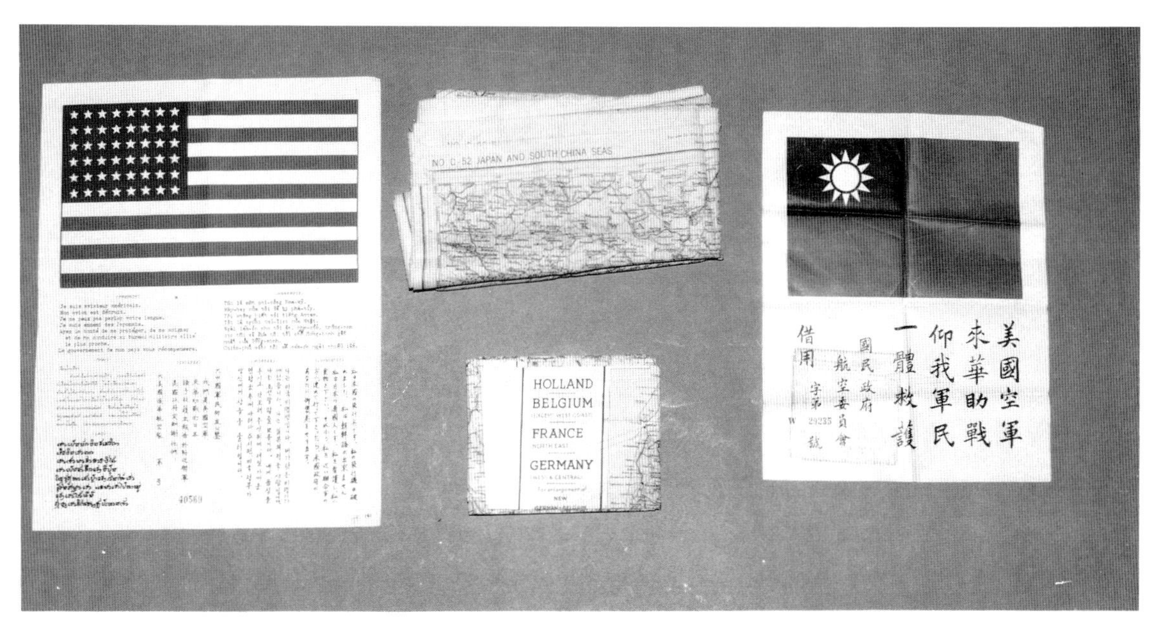

(Left): WWII blood chit, Asia (middle): escape maps (right): CBI blood chit

'Blood Chits' originated with the Flying Tigers in the CBI theater and are still being used. Each chit is serial numbered, assigned to an individual airman, and promises a reward or payment for the safe return of the holder. Blood chits are issued in the theater of operation and several local languages (besides English) are on each chit. Chits have been printed on silk, cloth and manmade fabrics.

Escape maps are issued for use in a specific geographic areas and many hundreds exist, essentially covering the major areas of the world. Escape maps can include survival and medical information and some are made of waterproof materials that allow their use as rain/sun shelters and water collectors.

Pointee Talkie language phrase books and miscellaneous language guides

BLOOD CHITS, ESCAPE MAPS, PHRASE BOOKS

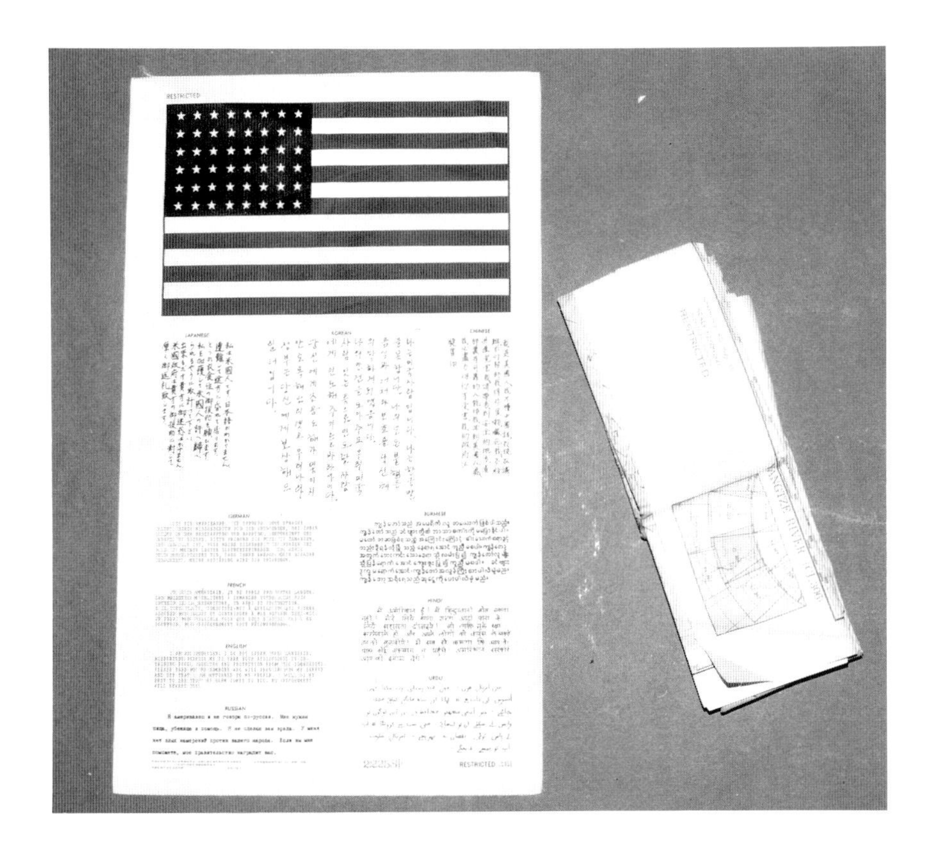

Korean war blood chit (left), escape map (right)

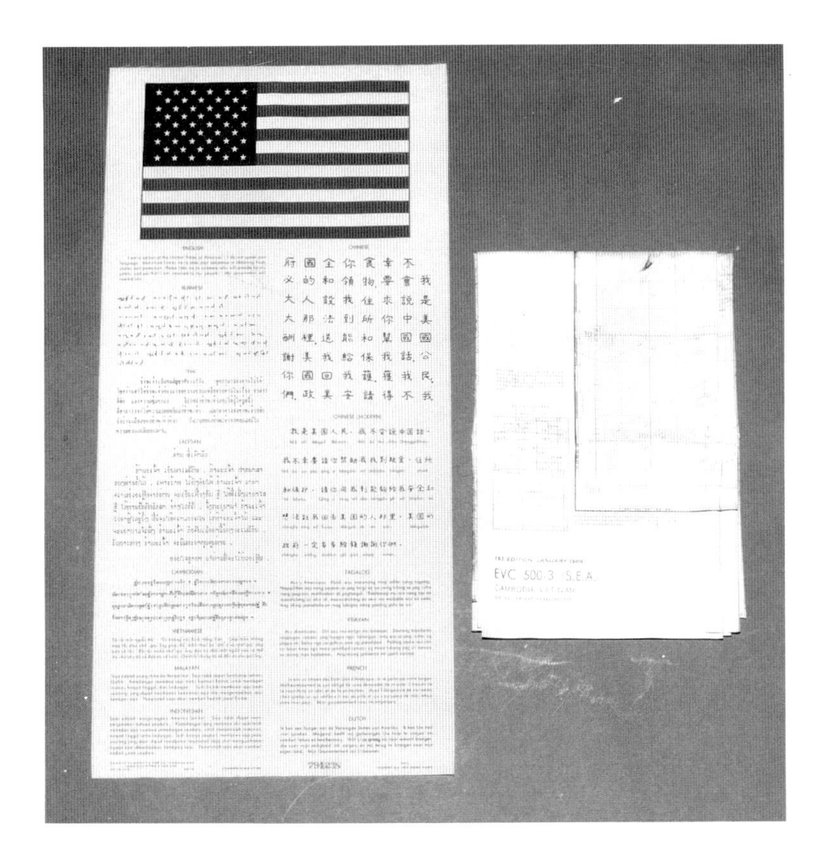

S.E.A. blood chit (left), Escape and Evasion Chart (right)

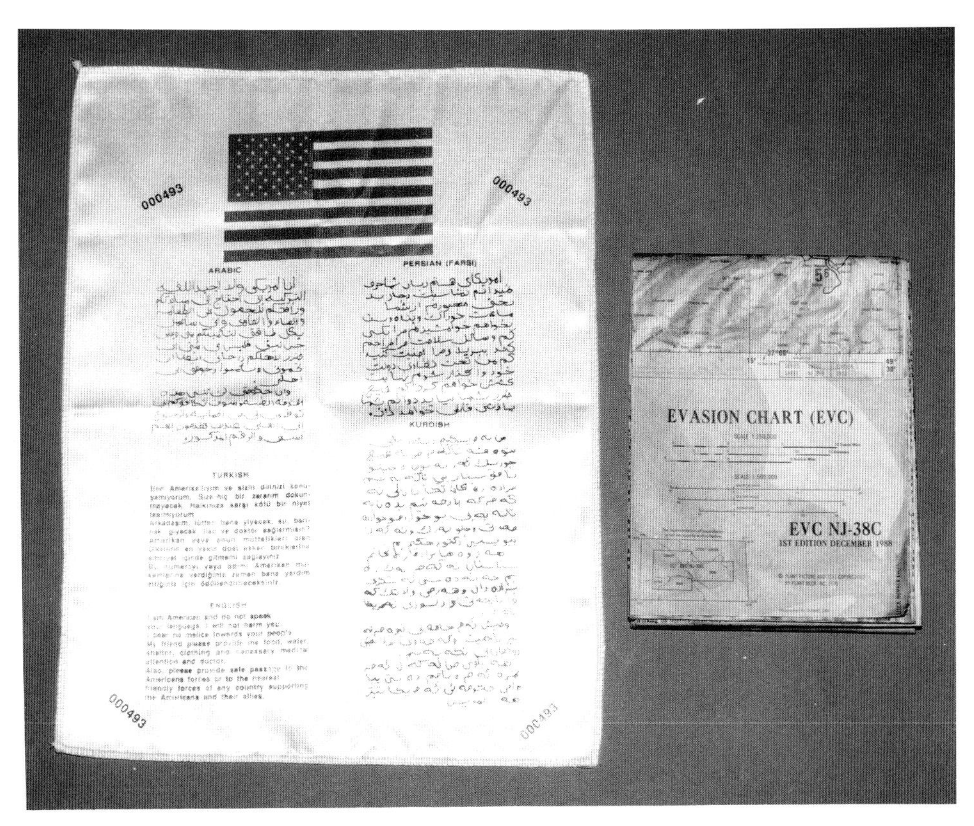

Desert Storm blood chit (left), Evasion chart (right)

APPENDIX

ASSEMBLY SHEET DRAWINGS

	PAGE
SRU-21/P VEST, U.S. ARMY	191
SRU-21/P VEST, U.S.A.F.	192
INDIVIDUAL OVERWATER KIT	193
INDIVIDUAL HOT CLIMATE KIT	194
INDIVIDUAL COLD CLIMATE KIT	195
OV-1 OVERWATER KIT	196
OV-1 HOT CLIMATE KIT	197
OV-1 COLD CLIMATE KIT	198

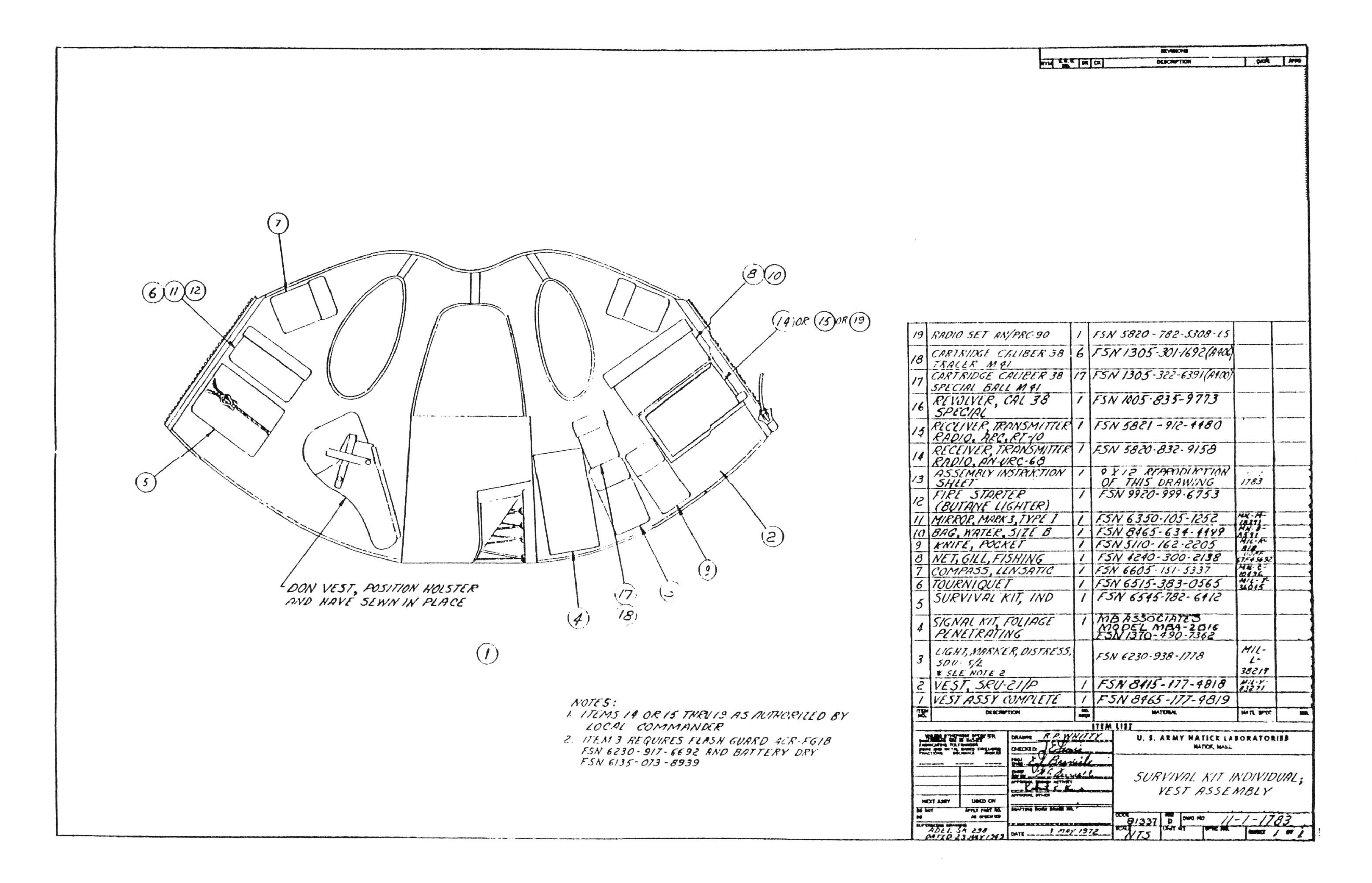

ITEM NO.	DESCRIPTION	NO. REQD	MATERIAL	MATL SPEC
19	RADIO SET AN/PRC-90	1	FSN 5820-782-5308-LS	
18	CARTRIDGE CALIBER 38 TRACER M41	6	FSN 1305-301-1692(A400)	
17	CARTRIDGE CALIBER 38 SPECIAL BALL M41	17	FSN 1305-322-6391(A400)	
16	REVOLVER, CAL 38 SPECIAL	1	FSN 1005-835-9773	
15	RECEIVER, TRANSMITTER RADIO, ARC, RT-10	1	FSN 5821-912-4480	
14	RECEIVER, TRANSMITTER RADIO, AN-URC-68	1	FSN 5820-832-9158	
13	ASSEMBLY INSTRUCTION SHEET	1	9 X 12 REPRODUCTION OF THIS DRAWING	1783
12	FIRE STARTER (BUTANE LIGHTER)	1	FSN 9920-999-6753	
11	MIRROR, MARK 3, TYPE I	1	FSN 6350-105-1252	[illegible]
10	BAG, WATER, SIZE B	1	FSN 8465-634-4499	[illegible]
9	KNIFE, POCKET	1	FSN 5110-162-2205	[illegible]
8	NET, GILL, FISHING	1	FSN 4240-300-2138	[illegible]
7	COMPASS, LENSATIC	1	FSN 6605-151-5337	[illegible]
6	TOURNIQUET	1	FSN 6515-383-0565	[illegible]
5	SURVIVAL KIT, IND	1	FSN 6545-782-6412	
4	SIGNAL KIT, FOLIAGE PENETRATING	1	MB ASSOCIATES MODEL MBA-2016 FSN 1370-490-7362	
3	LIGHT, MARKER, DISTRESS, SDU-5/E * SEE NOTE 2		FSN 6230-938-1778	MIL-L-38219
2	VEST, SRU-21/P	1	FSN 8415-177-4818	MIL-V-83271
1	VEST ASSY COMPLETE	1	FSN 8465-177-9819	

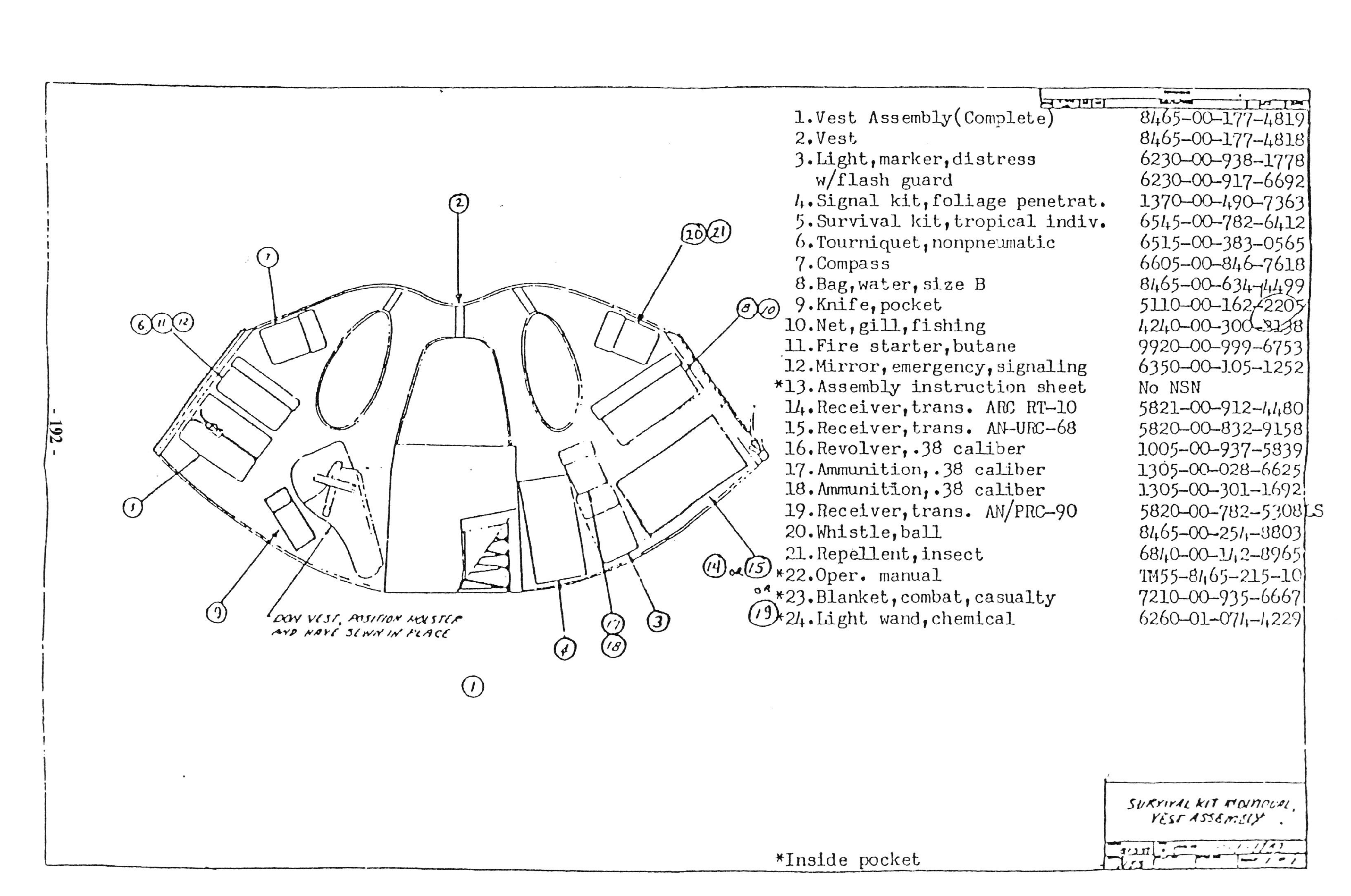

1.Vest Assembly(Complete) 8465-00-177-4819
2.Vest 8465-00-177-4818
3.Light,marker,distress 6230-00-938-1778
w/flash guard 6230-00-917-6692
4.Signal kit,foliage penetrat. 1370-00-490-7363
5.Survival kit,tropical indiv. 6545-00-782-6412
6.Tourniquet,nonpneumatic 6515-00-383-0565
7.Compass 6605-00-846-7618
8.Bag,water,size B 8465-00-634-4499
9.Knife,pocket 5110-00-162-2205
10.Net,gill,fishing 4240-00-300-3138
11.Fire starter,butane 9920-00-999-6753
12.Mirror,emergency,signaling 6350-00-105-1252
*13.Assembly instruction sheet No NSN
14.Receiver,trans. ARC RT-10 5821-00-912-4480
15.Receiver,trans. AN-URC-68 5820-00-832-9158
16.Revolver,.38 caliber 1005-00-937-5839
17.Ammunition,.38 caliber 1305-00-028-6625
18.Ammunition,.38 caliber 1305-00-301-1692
19.Receiver,trans. AN/PRC-90 5820-00-782-5308
20.Whistle,ball 8465-00-254-8803
21.Repellent,insect 6840-00-142-8965
*22.Oper. manual TM55-8465-215-10
*23.Blanket,combat,casualty 7210-00-935-6667
*24.Light wand,chemical 6260-01-074-4229
*Inside pocket
DON VEST, POSITION HOLSTER AND HAVE SEWN IN PLACE
SURVIVAL KIT INDIVIDUAL, VEST ASSEMBLY

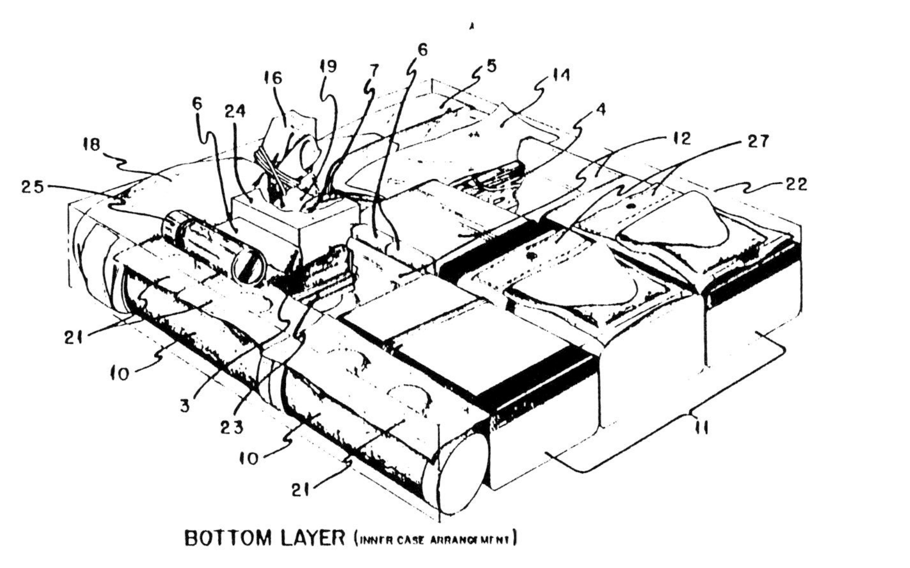

TOP LAYER (INNER-CASE ARRANGEMENT)

BOTTOM LAYER (INNER CASE ARRANGEMENT)

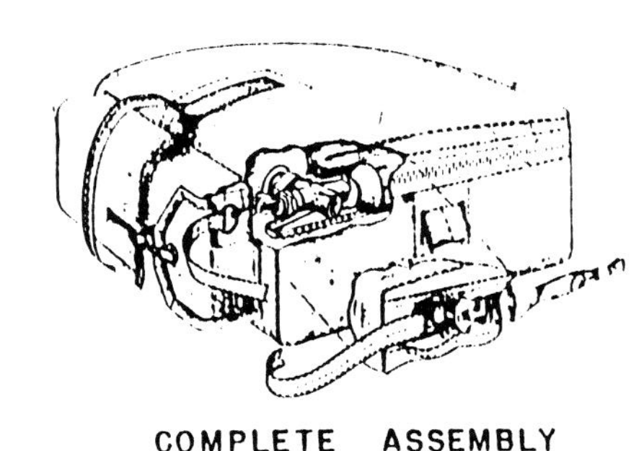

COMPLETE ASSEMBLY

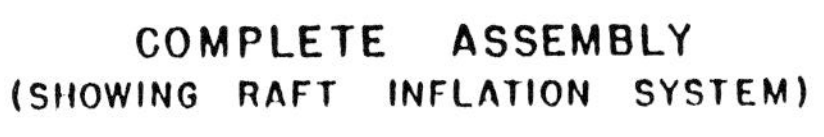

COMPLETE ASSEMBLY
(SHOWING RAFT INFLATION SYSTEM)

ITEM NO.	DESCRIPTION	QTY	PART OR STOCK NO
1	CASE, OUTER	1	8465-082-2511
2	CASE, INNER	1	8465-082-2514
3	FISHING TACKLE KIT	1	7810-558-2685
4	KNIFE, POCKET	1	5110-162-2205
5	FIRST AID KIT, AVIATOR, CAMOUFLAGED	1	6545-965-2394
6	MATCH, (NON-SAFETY, WOOD)	3	9920-985-6891
7	SUNBURN-PREVENTIVE PREPARATION	1	8510-162-5658
8	HAT AND MOSQUITO NET	1	8415-261-6630
9	HAT, REVERSIBLE, SUN	1	8415-270-0229
10	SIGNAL, SMOKE AND ILLUMINATION	2	1370-309-5028
11	DESALTER KIT, SEAWATER, MK 2	3	4220-216-5031
12	FOOD PACKET, SURVIVAL	3	8970-082-5665
13	BAG, STORAGE, WATER	1	8465-485-3034
14	SPOON, PLASTIC (FOR RATIONS)	1	7340-170-8374
15	LIFE RAFT, INFLATABLE, LRU - 3/P	1	4220-726-0424

ITEM NO.	DESCRIPTION	QTY	PART OR STOCK NO
16	BAILER, BOAT	1	2090-277-6583
17	PADDLE, BOAT (LIFE RAFT)	2	2040-485-3018
18	REPAIR KIT, INFLATABLE CRAFT	1	2090-693-1471
19	SPONGE, CELLULOSE, TYPE II	1	7920-240-2555
20	MANUAL, SURVIVAL FM 21-76	1	NSN
21	FUEL, COMPRESSED, TRIOXANE	3	9110-263-9865
22	PAN, FRYING	1	7330-082-2398
23	MIRROR, SIGNALING	1	6350-299-6197
24	COMPASS, MAGNETIC, TYPE MC-1, UNMOUNTED	1	6605-515-5637
25	BOX, MATCH, WATERPROOF	1	8465-265-4925
26	PACKING LIST (9x12 INCH)	1	QMC 11-1-172
27	SEA MARKER, FLUORESCEIN	2	6850-270-9986

SURVIVAL KIT
OVER WATER

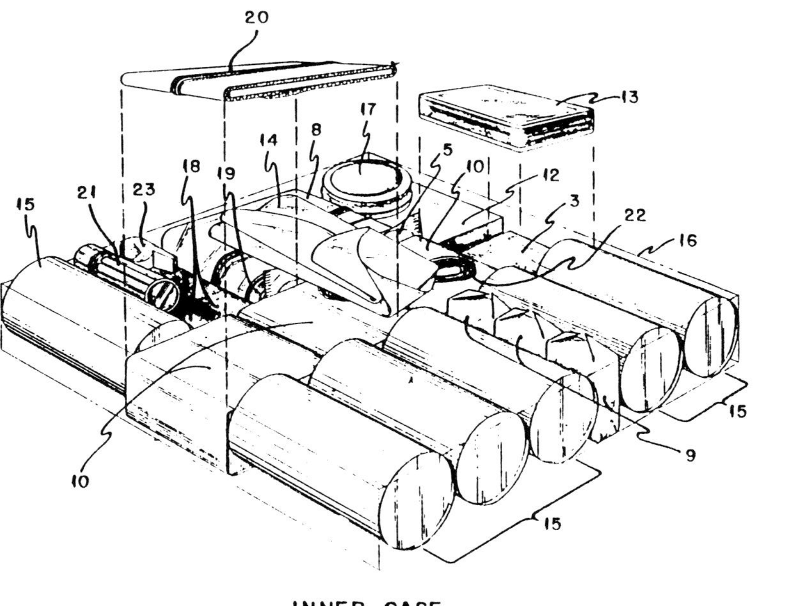

INNER CASE

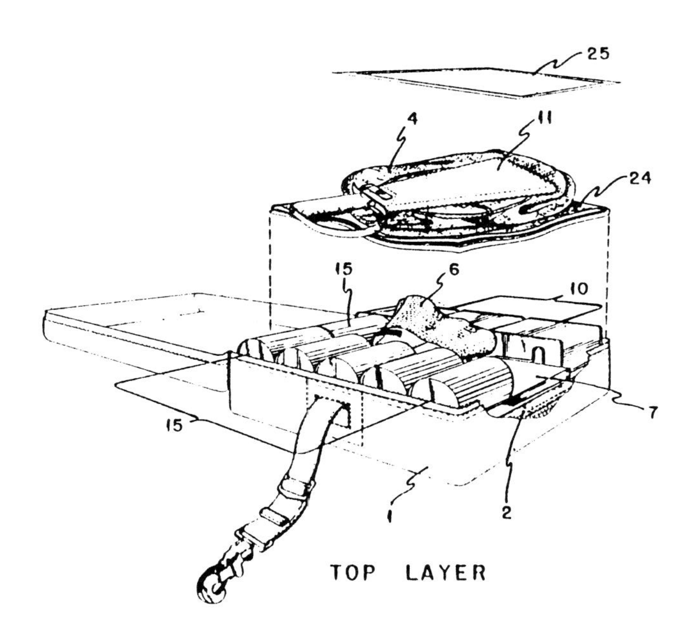

TOP LAYER

ITEM NO.	DESCRIPTION	QTY	PART OR STOCK NO
1	Case, Outer, Ind. Survival Kit	1	8465-082-2513
2	Case, Inner, Ind. Survival Kit	1	8465-082-2514
3	Compass, Magnetic, Type MC-1, Unmouted	1	6605-515-5637
4	Hat and Mosquito Net	1	8415-261-6630
5	Spoon, Plastic (For Rations)	1	7340-170-8374
6	Hat, Reversible, Sun	1	8415-270-0229
7	Manual, Survival AF 64-5	1	NSN
8	First Aid Kit, Aviator	1	6545-965-2394
9	Match, (Non-Safety)	3	9920-985-5891
10	Food Packet, Survival	6	8970-082-5665
11	Tool Kit, Survival	1	8465-973-4807
12	Fuel, Compressed, Trioxane	1	9110-263-9865
13	Fishing Tackle Kit	1	7810-558-2685
14	Bag, Storage, Drinking Water	1	8465-485-3034
15	Water, Drinking, Canned	12	8960-243-2103
16	Pan, Frying	1	7330-082-2398
17	Sunburn-Preventive Preparation	1	8510-162-5658
18	Signal, Smoke and Illumination, Marine MK 13, MOD 0	2	1370-309-5028

ITEM NO.	DESCRIPTION	QTY	PART OR STOCK NO.
19	Knife, Pocket	1	5110-162-2205
20	Mirror, Signaling, Emergency, MK 3	1	6350-299 6197.
21	Box, Match	1	8465-265-4925
22	Wire, Snare, 20 ft Long	1	9525-596-3498
23	Whistle, Ball, Plastic	1	8465-254-8803
24	Tarpaulin, Lightweight, Size 77	1	8340-485-3012
25	Packing List (9 x 12 Inch)	1	QMC 11-1-168

SURVIVAL KIT
HOT CLIMATE

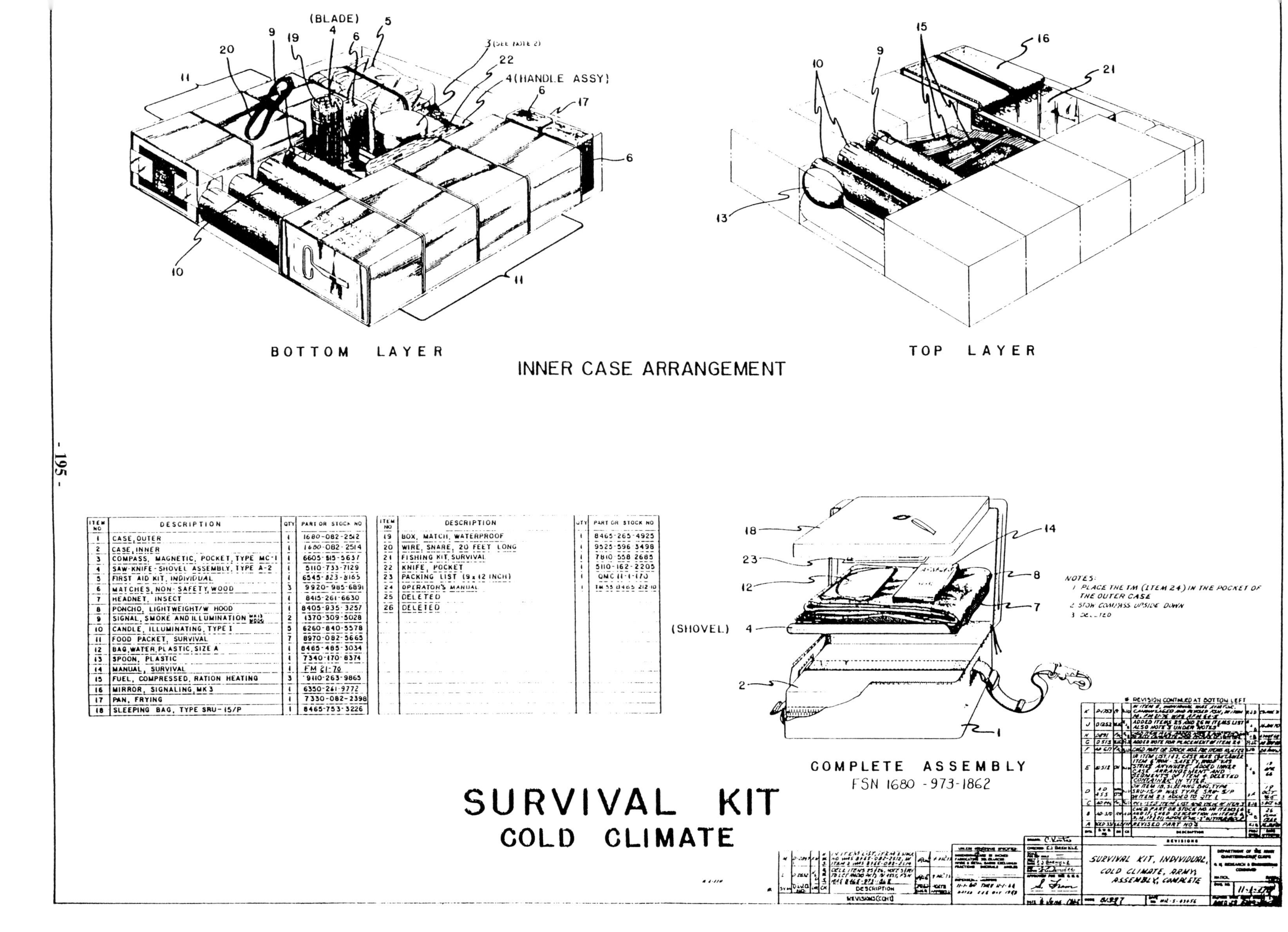

ITEM NO	DESCRIPTION	QTY	PART OR STOCK NO
1	CASE, OUTER	1	1680-082-2512
2	CASE, INNER	1	1680-082-2514
3	COMPASS, MAGNETIC, POCKET, TYPE MC-1	1	6605-815-5637
4	SAW-KNIFE-SHOVEL ASSEMBLY, TYPE A-2	1	5110-733-7129
5	FIRST AID KIT, INDIVIDUAL	1	6545-823-8165
6	MATCHES, NON-SAFETY, WOOD	3	9920-985-6891
7	HEADNET, INSECT	1	8415-261-6630
8	PONCHO, LIGHTWEIGHT/W HOOD	1	8405-935-3257
9	SIGNAL, SMOKE AND ILLUMINATION MK13 MOD0	2	1370-309-5028
10	CANDLE, ILLUMINATING, TYPE I	5	6260-840-5578
11	FOOD PACKET, SURVIVAL	7	8970-082-5665
12	BAG, WATER, PLASTIC, SIZE A	1	8465-485-3034
13	SPOON, PLASTIC	1	7340-170-8374
14	MANUAL, SURVIVAL	1	FM 21-76
15	FUEL, COMPRESSED, RATION HEATING	3	9110-263-9865
16	MIRROR, SIGNALING, MK3	1	6350-261-9772
17	PAN, FRYING	1	7330-082-2398
18	SLEEPING BAG, TYPE SRU-15/P	1	8465-753-3226

ITEM NO	DESCRIPTION	QTY	PART OR STOCK NO
19	BOX, MATCH, WATERPROOF	1	8465-265-4925
20	WIRE, SNARE, 20 FEET LONG	1	9525-596-3498
21	FISHING KIT, SURVIVAL	1	7810-558-2685
22	KNIFE, POCKET	1	5110-162-2205
23	PACKING LIST (9 x 12 INCH)	1	QMC 11-1-170
24	OPERATOR'S MANUAL	1	TM 55-8465-212-10
25	DELETED		
26	DELETED		

SURVIVAL KIT
COLD CLIMATE

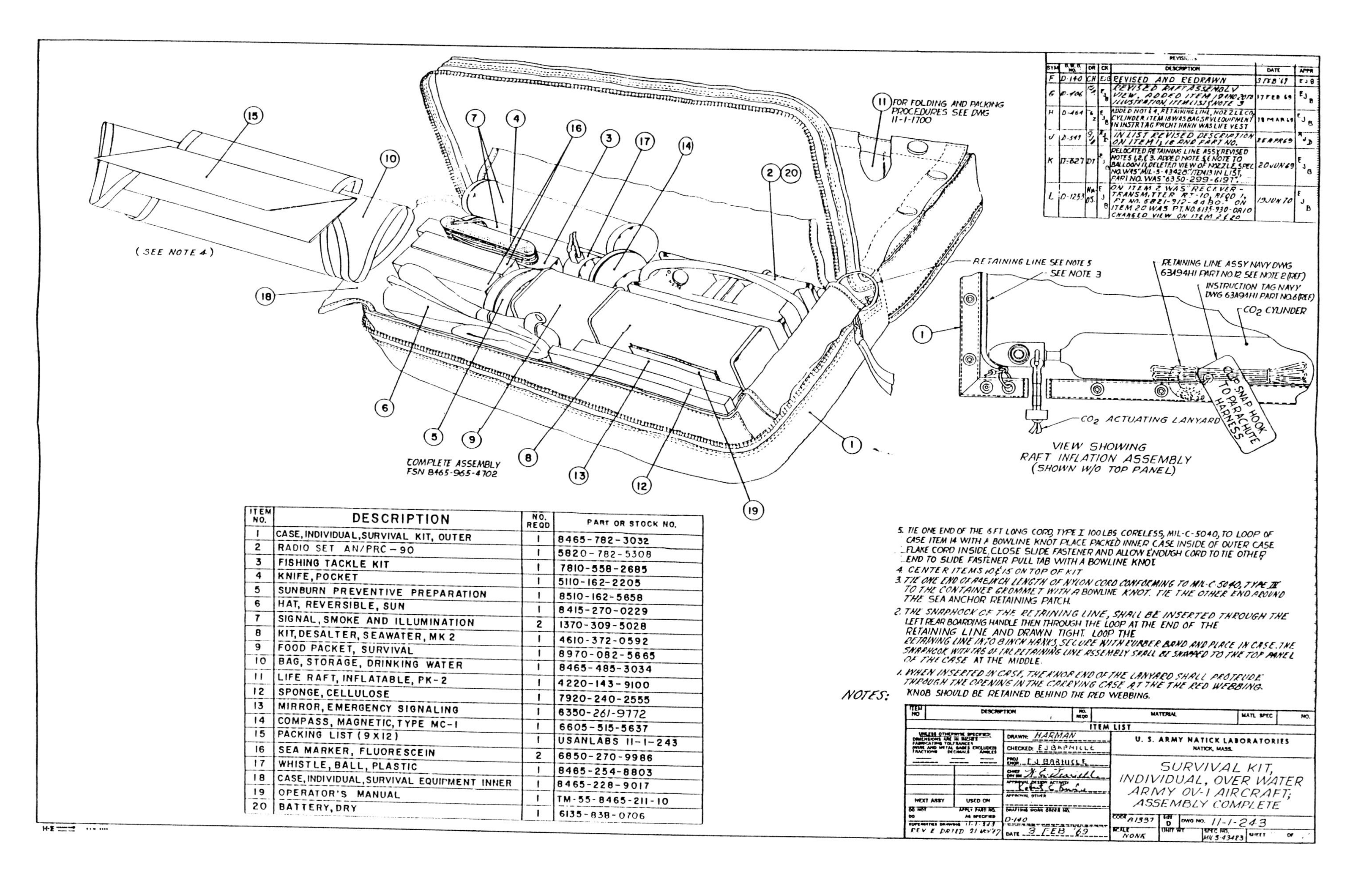

ITEM NO.	DESCRIPTION	NO. REQD	PART OR STOCK NO.
1	CASE, INDIVIDUAL, SURVIVAL KIT, OUTER	1	8465-782-3032
2	RADIO SET AN/PRC-90	1	5820-782-5308
3	FISHING TACKLE KIT	1	7810-558-2685
4	KNIFE, POCKET	1	5110-162-2205
5	SUNBURN PREVENTIVE PREPARATION	1	8510-162-5658
6	HAT, REVERSIBLE, SUN	1	8415-270-0229
7	SIGNAL, SMOKE AND ILLUMINATION	2	1370-309-5028
8	KIT, DESALTER, SEAWATER, MK 2	1	4610-372-0592
9	FOOD PACKET, SURVIVAL	1	8970-082-5665
10	BAG, STORAGE, DRINKING WATER	1	8465-485-3034
11	LIFE RAFT, INFLATABLE, PK-2	1	4220-143-9100
12	SPONGE, CELLULOSE	1	7920-240-2555
13	MIRROR, EMERGENCY SIGNALING	1	6350-261-9772
14	COMPASS, MAGNETIC, TYPE MC-1	1	6605-515-5637
15	PACKING LIST (9X12)	1	USANLABS 11-1-243
16	SEA MARKER, FLUORESCEIN	2	6850-270-9986
17	WHISTLE, BALL, PLASTIC	1	8465-254-8803
18	CASE, INDIVIDUAL, SURVIVAL EQUIPMENT INNER	1	8465-228-9017
19	OPERATOR'S MANUAL	1	TM-55-8465-211-10
20	BATTERY, DRY	1	6135-838-0706

NOTES:

5. TIE ONE END OF THE 6 FT LONG CORD, TYPE I 100 LBS CORELESS, MIL-C-5040, TO LOOP OF CASE ITEM 14 WITH A BOWLINE KNOT. PLACE PACKED INNER CASE INSIDE OF OUTER CASE. FLAKE CORD INSIDE, CLOSE SLIDE FASTENER AND ALLOW ENOUGH CORD TO TIE OTHER END TO SLIDE FASTENER PULL TAB WITH A BOWLINE KNOT.

4. CENTER ITEMS 10 & 15 ON TOP OF KIT.

3. TIE ONE END OF 8 INCH LENGTH OF NYLON CORD CONFORMING TO MIL-C-5040, TYPE III TO THE CONTAINER GROMMET WITH A BOWLINE KNOT. TIE THE OTHER END AROUND THE SEA ANCHOR RETAINING PATCH.

2. THE SNAPHOOK OF THE RETAINING LINE, SHALL BE INSERTED THROUGH THE LEFT REAR BOARDING HANDLE THEN THROUGH THE LOOP AT THE END OF THE RETAINING LINE AND DRAWN TIGHT. LOOP THE RETAINING LINE INTO 8 INCH HANKS, SECURE WITH RUBBER BAND AND PLACE IN CASE. THE SNAPHOOK WITH TAG OF THE RETAINING LINE ASSEMBLY SHALL BE SNAPPED TO THE TOP PANEL OF THE CASE AT THE MIDDLE.

1. WHEN INSERTED IN CASE, THE KNOB END OF THE LANYARD SHALL PROTRUDE THROUGH THE OPENING IN THE CARRYING CASE AT THE THE RED WEBBING. KNOB SHOULD BE RETAINED BEHIND THE RED WEBBING.

REVISIONS

SYM	E.W.R. NO.	DR	CK	DESCRIPTION	DATE	APPR
F	D-140	CH	E.G.	REVISED AND REDRAWN	3 FEB 69	EJB
G	D-406			REVISED RAFT ASSEMBLY VIEW. ADDED ITEM 19 AND 20 TO ILLUSTRATION, ITEM 15 SEE NOTE 3	17 FEB 69	EJB
H	D-464			ADDED NOTE 4, RETAINING LINE, NOZZLE CO_2 CYLINDER. ITEM 18 WAS BAG, SRVL EQUIPMENT IN INSTR TAG PRCHT HARN WAS LIFE VEST	18 MAR 69	JB
J	D-511			IN LIST REVISED DESCRIPTION ON ITEM 1, 18 AND PART NO.	28 APR 69	JB
K	D-827	DT		RELOCATED RETAINING LINE ASSY. REVISED NOTES 1, 2 & 3. ADDED NOTE 5 & NOTE TO BALLOON 11. DELETED VIEW OF NOZZLE. SPEC NO. WAS MIL-S-43420. ITEM 13 IN LIST, PART NO. WAS 6350-299-6197.	20 JUN 69	EJB
L	D-1253			ON ITEM 2 WAS RECEIVER-TRANSMITTER RT-10, REQD 1, PT NO. 5821-912-4480. ON ITEM 20 WAS PT. NO. 6135-930-0810. CHANGED VIEW ON ITEM 2 & 20	19 JUN 70	EJB

ITEM LIST

U. S. ARMY NATICK LABORATORIES
NATICK, MASS.

DRAWN: HARMAN
CHECKED: E J BARNICLE
PROJ ENGR: E J BARNICLE

SURVIVAL KIT, INDIVIDUAL, OVER WATER ARMY OV-1 AIRCRAFT; ASSEMBLY COMPLETE

CODE 81337 — SIZE D — DWG NO. 11-1-243
SCALE NONE
SPEC NO. MIL-S-43483
D-140 — DATE 3 FEB 69

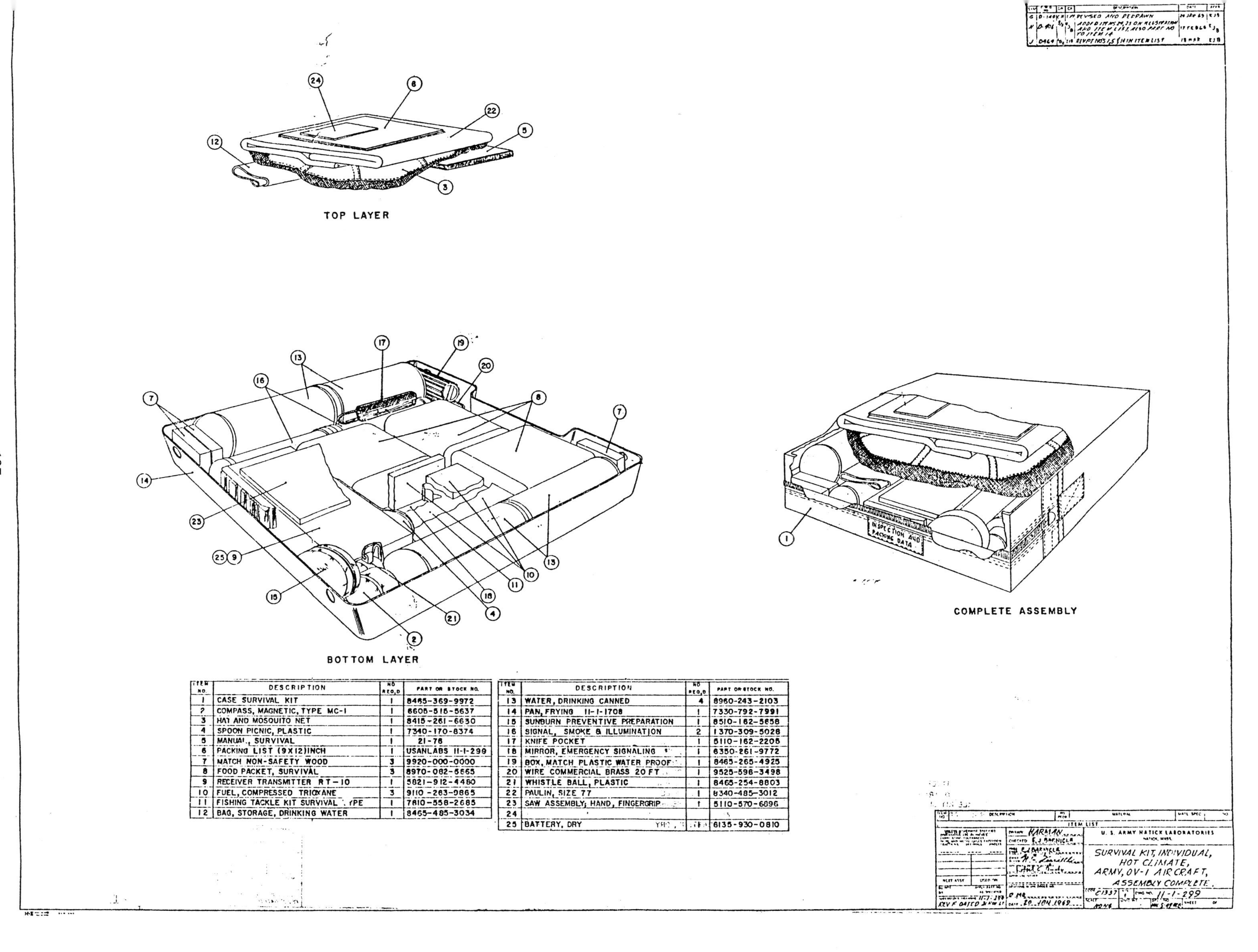

ITEM NO.	DESCRIPTION	NO REQ'D	PART OR STOCK NO.
1	CASE SURVIVAL KIT	1	8465-369-9972
2	COMPASS, MAGNETIC, TYPE MC-1	1	6605-515-5637
3	HAT AND MOSQUITO NET	1	8415-261-6630
4	SPOON PICNIC, PLASTIC	1	7340-170-8374
5	MANUAL, SURVIVAL	1	21-76
6	PACKING LIST (9X12)INCH	1	USANLABS 11-1-299
7	MATCH NON-SAFETY WOOD	3	9920-000-0000
8	FOOD PACKET, SURVIVAL	3	8970-082-5665
9	RECEIVER TRANSMITTER RT-10	1	5821-912-4480
10	FUEL, COMPRESSED TRIOXANE	3	9110-263-9865
11	FISHING TACKLE KIT SURVIVAL TYPE	1	7810-558-2685
12	BAG, STORAGE, DRINKING WATER	1	8465-485-3034
13	WATER, DRINKING CANNED	4	8960-243-2103
14	PAN, FRYING 11-1-1708	1	7330-792-7991
15	SUNBURN PREVENTIVE PREPARATION	1	8510-162-5658
16	SIGNAL, SMOKE & ILLUMINATION	2	1370-309-5028
17	KNIFE POCKET	1	5110-162-2205
18	MIRROR, EMERGENCY SIGNALING	1	6350-261-9772
19	BOX, MATCH PLASTIC WATER PROOF	1	8465-265-4925
20	WIRE COMMERCIAL BRASS 20 FT	1	9525-596-3498
21	WHISTLE BALL, PLASTIC	1	8465-254-8803
22	PAULIN, SIZE 77	1	8340-485-3012
23	SAW ASSEMBLY, HAND, FINGERGRIP	1	5110-570-6096
24			
25	BATTERY, DRY		6135-930-0810

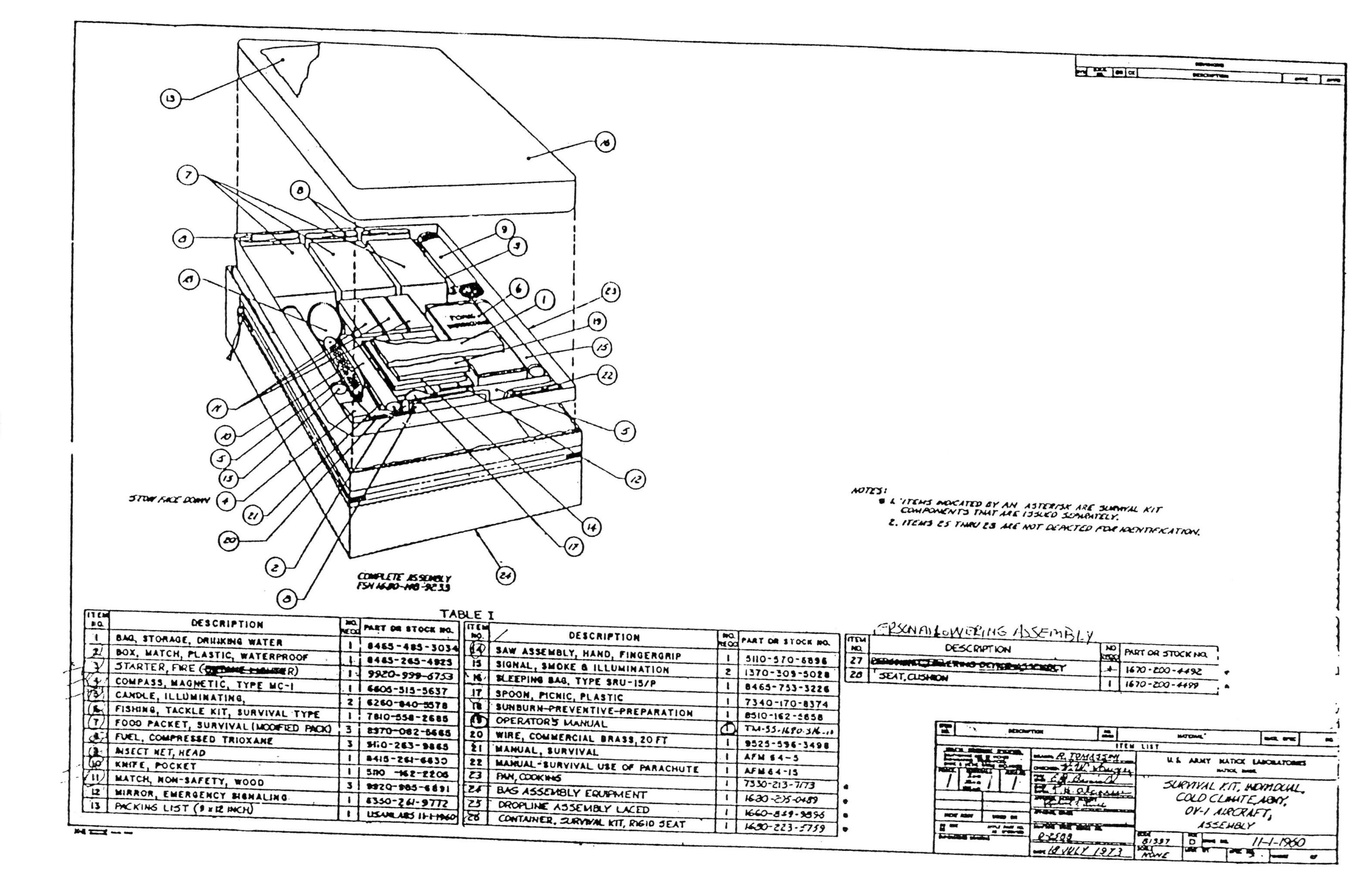

TABLE I

ITEM NO.	DESCRIPTION	NO. REQD	PART OR STOCK NO.
1	BAG, STORAGE, DRINKING WATER	1	8465-485-3034
2	BOX, MATCH, PLASTIC, WATERPROOF	1	8465-265-4923
3	STARTER, FIRE (R)	1	9920-999-6753
4	COMPASS, MAGNETIC, TYPE MC-1	1	6605-515-5637
5	CANDLE, ILLUMINATING,	2	6260-840-5578
6	FISHING, TACKLE KIT, SURVIVAL TYPE	1	7810-558-2685
7	FOOD PACKET, SURVIVAL (MODIFIED PACK)	3	8970-082-5665
8	FUEL, COMPRESSED TRIOXANE	3	9110-263-9865
9	INSECT NET, HEAD	1	8415-261-6630
10	KNIFE, POCKET	1	5110-162-2206
11	MATCH, NON-SAFETY, WOOD	3	9920-985-6891
12	MIRROR, EMERGENCY SIGNALING	1	6350-261-9772
13	PACKING LIST (9 x 12 INCH)	1	USANLABS 11-1-1960
14	SAW ASSEMBLY, HAND, FINGERGRIP	1	5110-570-6896
15	SIGNAL, SMOKE & ILLUMINATION	2	1370-309-5028
16	SLEEPING BAG, TYPE SRU-15/P	1	8465-753-3226
17	SPOON, PICNIC, PLASTIC	1	7340-170-8374
18	SUNBURN-PREVENTIVE-PREPARATION	1	8510-162-5658
19	OPERATOR'S MANUAL	1	TM-55-1680-316-10
20	WIRE, COMMERCIAL BRASS, 20 FT	1	9525-596-3498
21	MANUAL, SURVIVAL	1	AFM 64-5
22	MANUAL-SURVIVAL USE OF PARACHUTE	1	AFM 64-15
23	PAN, COOKING	1	7330-213-7173 *
24	BAG ASSEMBLY EQUIPMENT	1	1630-295-0489 *
25	DROPLINE ASSEMBLY LACED	1	1660-859-9896 *
26	CONTAINER, SURVIVAL KIT, RIGID SEAT	1	1630-223-5759 *

PERSONNEL LOWERING ASSEMBLY

ITEM NO.	DESCRIPTION	NO. REQD	PART OR STOCK NO.
27		1	1670-200-4492 *
28	SEAT, CUSHION	1	1670-200-4499 *

REFERENCES

USAF SUPPLY CATALOG
CLASS 20-B
SURVIVAL EQUIPMENT AND PARACHUTES
19 MAY 1951 (REV 19 NOV 1951)

NAVAER 00-80T-56
SURVIVAL TRAINING GUIDE
USN
NOV. 1955

TM 11-5820-640-15
DEPT OF THE ARMY TECHNICAL MANUAL
RADIO SETS AN/URC-10, AN/URC-10A, & ACR RT-10
MAY 1967

TM 11-5820-767-12 DEPT OF THE ARMY TECHNICAL MANUAL
RADIO SET AN/URC-68
AUG 1968

NAVAER 00-80T-52
SAFETY AND SURVIVAL EQUIPMENT FOR NAVAL AVIATION
1959

AFM 64-4
HANDBOOK OF PROTECTIVE EQUIPMENT
USAF
OCT 1954, 30 MARCH 1964, 29 NOV 1974

SC 8465-90-CL-P02
DEPT OF THE ARMY SUPPLY CATALOG
SURVIVAL KIT, INDIVIDUAL, VEST TYPE
15 JULY 1974

AIRCREW SURVIVAL EQUIPMENTMAN 3 & 2, VOL 1

AF T.O. 14D3-11-1

TM 55-8465-215-10
DEPT OF THE ARMY TECHNICAL MANUAL
OPERATORS MANUAL FOR VEST, SURVIVAL, SRU 21/P, HOT CLIMATE
JUNE 1970

TM 55-1680-316-10
TECHNICAL MANUAL DEPT OF THE ARMY
OPERATORS MANUAL FOR RIGID SEAT SURVIVAL KIT AND SURVIVAL VEST
OV-1 AIRCRAFT

TM 55-1680-351-10
DEPT. OF THE ARMY TECHNICAL MANUAL
OPERATORS MANUAL FOR SRU-21/P, ARMY VEST, SURVIVAL
22 APRIL 1987

TM 55-8465-214-10
TECHNICAL MANUAL DEPT OF THE ARMY
OPERATORS MANUAL, SURVIVAL KIT, OVERWATER, INDIVIDUAL
AUG 1971

TM 55-8464-212-10
TECHNICAL MANUAL DEPT OF THE ARMY
OPERATORS MANUAL, SURVIVAL KIT, COLD CLIMATE, INDIVIDUAL
JUNE 1971

REFERENCES

TM 55-8465-213-10
TECHNICAL MANUAL DEPT OF THE ARMY
OPERATORS MANUAL, SURVIVAL KIT, HOT CLIMATE, INDIVIDUAL
JUNE 1971

MOTOROLA, GOVT ELECTRONICS GROUP
COMMUNICATIONS DIV
COMMUNICATIONS PRODUCTS OFFICE
8201 E. MCDOWELL RD P.O. BOX 1417
SCOTTSDALE, AZ 85252-1417
AN/PRC-112 RADIO

US ARMY NATICK RESEARCH, DEVELOPMENT AND ENGINEERING CENTER
NATICK, MA 01760
CAPTAIN SHEILA M. RYAN
COMBAT ARMS PROJECT OFFICE/LIAISON DIV
SARVIP SURVIVAL VEST

TB ORD 409
DEPT OF THE ARMY TECHNICAL BULLETIN
CAL. .22/.410 GAGE SURVIVAL RIFLE-SHOTGUN M6

TB ORD 389
DEPT OF THE ARMY TECHNICAL BULLETIN
CAL. .22 SURVIVAL RIFLE M4
(HORNET CARTRIDGE) (T38)

ARMY AIRCRAFT SURVIVAL KITS
TM 55-1680-317-23 & P
26 MARCH 1987

T.O. 04-15-1A
OPERATION AND SERVICE INSTRUCTIONS WITH ACCESSORY PARTS LIST-
LIFE RAFTS
30 OCT 1950
REV. T.O. 1453-1-31 15 APR 1955

T.O. 14-S10-2-2
TM 55-1680-322-12
OPERATION & SERVICE DISTRESS MARKER LIGHT PART NO. SDU 5/E

UNITED STATES NAVY (BUREAU OF AERONAUTICS)
AIR-SEA RESCUE EQUIPMENT WWII

NAVAIR 13-1-6.3 RIGID SEAT SURVIVAL KIT - 8 SERIES
NAVAIR 13-1-6.5
NAVAIR 13-1-6.7 CMU-24/P SURVIVAL VEST

NAVWEPS 13-5-501
(USAF) T.O. 14D1-2-281
PARACHUTE MANUAL
1 JULY 1963 (AF REPRINT 29 NOV 74)

TECHNICAL ORDER NO. 13-5-46
PARACHUTE - ATTACHMENT OF TYPES B-2, B-4, AND C-1 PARACHUTE
EMERGENCY KITS
5 SEPTEMBER 1944

SC 8454-90-CL-P03
DEPT. OF THE ARMY SUPPLY CATALOG
SURVIVAL KIT, INDIVIDUAL, COLD CLIMATE
OCT 1971

REFERENCES

USAF T.O. 14S1-3-71
AND T.O. 14S1-3-51 AIRFORCE SURVIVAL EQUIPMENT

TM 55-1680-317-23 & P
TECHNICAL MANUAL
ARMY AIRCRAFT SURVIVAL KITS
8 AUG 1975 AND 26 MARCH 1987

TM 11-5820-800-12
TECHNICAL MANUAL
RADIO SET AN/PRC-90
NOV 1973

AIRFORCE T.O. 14S-1-102
ARMY TM 5-4220-202-14
USAF FLOATATION EQUIPMENT
JUNE 1976

US ARMY AIRFORCES
CLASS 13 ILLUSTRATED CATALOG - CLOTHING, PARACHUTES , EQUIPMENT
AND SUPPLIES
SEPT 30, 1943 (REV. 1 APRIL 1944)

AAF MANUAL 55-0-1
REFERENCE MANUAL FOR PERSONAL EQUIPMENT OFFICERS
JUNE 1945

AIRFORCE T.O. 14S3-1-3
TYPES & NUMBERS OF INDIVIDUAL SURVIVAL KIT CONTAINERS AND LIFE RAFTS
TO BE USED IN VARIOUS TYPE AIRCRAFT
31 MAY 1968